SpringerBriefs in Applied Sciences and Technology

PoliMI SpringerBriefs

More information about this series at http://www.springer.com/series/11159
http://www.polimi.it

Laura Cattaneo · Sergio Terzi
Editors

Models, Methods and Tools for Product Service Design

The Manutelligence Project

POLITECNICO
MILANO 1863

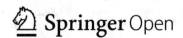

Springer Open

Editors
Laura Cattaneo
Department of Management, Economics
  and Industrial Engineering
Politecnico di Milano
Milan, Italy

Sergio Terzi
Department of Management, Economics
  and Industrial Engineering
Politecnico di Milano
Milan, Italy

ISSN 2191-530X   ISSN 2191-5318   (electronic)
SpringerBriefs in Applied Sciences and Technology
ISSN 2282-2577   ISSN 2282-2585   (electronic)
PoliMI SpringerBriefs
ISBN 978-3-319-95848-4   ISBN 978-3-319-95849-1   (eBook)
https://doi.org/10.1007/978-3-319-95849-1

Library of Congress Control Number: 2018948601

This Springer imprint is published by the registered company Springer Nature Switzerland AG
The registered company address is: Gewerbestrasse 11, 6330 Cham, Switzerland

Preface

This book summarizes work being undertaken within the Manutelligence European Research Project (Grant agreement N°: 636951, H2020-FoF-2014, FoF-05—Innovative product–service design using manufacturing intelligence). The project aims at supporting enterprises to develop smart, social, and flexible products with high value-added services. Manutelligence has improved product and service design by developing suitable models and methods, and connecting them through a modular collaborative secure ICT Platform. The use of real data collected in real time through the Internet of Things (IoT) technologies underpin the design of product–service system (PSS) and allows to follow the PSS along its life cycle. Available data allows a better measure and simulation of costs and sustainability issues, through life-cycle cost (LCC) and life-cycle assessment (LCA). Analyzing data coming from IoT systems and sharing LCC and LCA information thanks to the ICT Platform allows speeding up the design of product–service (P-S), decreasing costs and better understanding customer needs. Industrial partners involved in the project provided a clear overview of the Manutelligence results and proved how its technological solutions improve the design of a product–service system and the management of the product–service life cycle.

The book covers a large number of topics, since Manutelligence really involved several issues coming from the product and service life cycle. It was designed to offer readers the possibility to have a complete view of all the results we have achieved during the project. Furthermore, it contains a clear explanation of the IT modular architecture we have developed in order to collect within a unique and complete framework different tools and software.

Chapter 1 introduces the main research contents and provides an overview on Manutelligence objectives. Furthermore, it introduces the description of the IT modular platform.

Chapter 2 deals with engineering and business requirements definition, analysis, and validation. It describes the four-phase methodology implemented to define the common aggregated requirements for the platform development. The phases include requirement elicitation, structuration and organization, analysis and refinement and validation. It also shows the main results of each phase.

Chapter 3 describes an approach to manage PSS along its life cycle. It includes a design methodology for PSS and a systems modeling method. It also highlights challenges related to PSS lifecycle management observed during the Manutelligence project.

Chapter 4 shows how the platform developed during the project enable designers and engineers to access through natural 3DEXPERIENCE to data from both the traditional enterprise IT systems (CAD, CAX, PLM, MES, etc.) and IoT-enabled systems. Furthermore, it describes how it is possible to retrieve physical products information and knowledge management during the PSS lifecycle phases.

Chapter 5 presents tools and procedures to embed and retrieve the knowledge related to the P-S and its life cycle in manufacturing. Taking into account the information coming from different sources (PLCs, sensorial IoT nodes, etc.) in the context of production system, a methodology to present coherently field data is proposed, with the idea to offer interoperability and integration also from the point of view of the different devices used to collect these data.

Chapter 6 describes a tool aimed at carrying out the life cycle assessment (called MaGA) and another one for the life cycle costing (called BAL.LCPA). In order to seamlessly include environmental and economic considerations into the design process, the two stand-alone tools have been integrated with the Manutelligence design platform. Their application in a Fablab-like environment is described to show how they interact with design tools and to provide examples of the results they get.

Chapter 7 describes all the different use cases involved in the project. It is focused on how each of them has applied Manutelligence methods and tools in order to improve PSS design and management. A particular focus is dedicated to the usage of the Manutelligence platform.

Chapter 8 illustrates the business potential of product and service lifecycle engineering tools within the manufacturing sector, with particular attention to the use cases of the Manutelligence project. The whole process (the aggregation and the value creation) is investigated. From the scenario, analysis of the global PLM sector emerged that the aggregation of product and service lifecycle engineering tools is able to generate significant added value, highlighting in this way the relevant commercial perspectives of the Manutelligence platform.

Milan, Italy   Laura Cattaneo
                                                                                      Sergio Terzi

Acknowledgements

We are grateful to our project officer Erastos Filos and to our reviewer Mrs. Anastasia Garbi for their constructive and useful suggestions; they greatly helped us shape our project results. Also, we own gratitude to our project coordinator Maurizio Petrucciani for his invaluable support through all phases of the project and to all the consortium partners, that have made this work and this project outstanding. At last, as authors from Politecnico di Milano we would like to thank the members of our team who were involved in the project, namely Francesca Amato, Daniele Cerri, Matteo Cocco, Elisabetta De Berti, Silvia Marchetto, Manuel Oliveira, Monica Rossi, Emanuela Vinci. All of them provided their best effort in achieving the results of the project, contributing to the research development and to the scientific improvement of our group, led by Prof. Marco Taisch, and to the progress of the entire Politecnico and to the whole society.

The work reported in this book is funded by the European Commission, through European Union's Horizon 2020 research and innovation programme, under grant agreement no. 636951.

Contents

Chapter 1
Introduction

Laura Cattaneo, Jacopo Cassina, Maurizio Petrucciani, Sergio Terzi and Stefan Wellsandt

Abstract This introductive chapter aims to clarify some of the main research contents that are involved in Manutelligence project and wants to present the objectives of the project and the structure of the Manutelligence IT platform. We briefly describe some fundamental concepts, such as the Product Lifecycle Management (PLM), the Product Service System (PSS), the Internet of Things (IoT) for the smart manufacturing, the Life Cycle Cost and the Life Cycle Assessment (LCC and LCA). All these topics are strictly connected, since Manutelligence project aims at supporting enterprises to design and to develop suitable Product-Service Systems, addressing customers' needs and stakeholders' requirements, collected also through IoT technologies. Furthermore it aims to integrate best in class methodology and tools from research and industry, resulting in a secure, collaborative Product/Service Design and Manufacturing Engineering Platform, able to manage the Product-Service lifecycle and to collect information in order to implement LCC and LCA.

L. Cattaneo (✉) · S. Terzi
Department of Economics, Management and Industrial Engineering, Politecnico di Milano,
20156 Milan, Italy
e-mail: laura1.cattaneo@polimi.it

S. Terzi
e-mail: sergio.terzi@polimi.it

S. Wellsandt
BIBA—Bremer Institut für Produktion und Logistik GmbH, University of Bremen,
Hochschulring 20, 28359 Bremen, Germany
e-mail: wel@biba.uni-bremen.de

M. Petrucciani
Dassault Systemes Italia Srl, Viale dell'Innovazione 3, 20126 Milano, Italy
e-mail: maurizio.petrucciani@3ds.com

J. Cassina
Holonix s.r.l, Corso Italia, 8, 20821 Meda, MB, Italy
e-mail: Jacopo.cassina@holonix.it

© The Author(s) 2019
L. Cattaneo and S. Terzi (eds.), *Models, Methods and Tools for Product Service Design*, PoliMI SpringerBriefs, https://doi.org/10.1007/978-3-319-95849-1_1

1.1 Product Lifecycle Management

PLM (Product Lifecycle Management) is an acronym widely used in the current industrial practice. Coined more than 15 years ago, PLM is often seen as an extensive and comprehensive concept, which defines the integration of different kinds of activities performed by engineering staff along the entire lifecycle of industrial products, "from cradle to grave" [1].

In its practical essence, PLM defines the adoption of several software tools and platforms for supporting innovation and engineering processes. According to the main business analysts (e.g. Gartner, CIMdata, Tech Clarity), PLM is a leading global market of IT solutions, mainly segmented in two branches: (i) Authoring and Simulation tools and (ii) Collaborative Product Development platforms and environments. In the first segment, dozens of vendors are globally proposing their solutions for enabling virtual prototyping solutions (from CAD 3D, to Computational Flow Dynamic, from Finite Element Analysis, to Discrete Event Simulation, etc.). The second branch is populated by a plethora of collaborative functionalities supporting, for instance, effective file sharing, document vaulting, work flow automation, team management and on distance working. Most of them are provided in one single, secured environment.

PLM is still a matter of design and engineering tools, and their integration. The industrial practice shows how PLM's real implementation is quite far from its comprehensive "lifecycle" meaning [2].

One product lifecycle framework in production engineering differentiates three main phases, describing the product from the "cradle to grave" [3]:

- **Beginning of Life (BOL)**: processes related to development, production and distribution;
- **Middle of Life (MOL)**: processes related to a product's use, service and repair;
- **End of Life (EOL)**: processes related to reverse logistics like reuse, recycle and disposal.

Approaches, such as closed-loop PLM [4], take a view upon the entire product lifecycle, from product ideation to end-of-life processes. Ideally, the view extends into the beginning of the next lifecycle. This puts forward a paradigm shift from "cradle to grave" to "cradle to cradle" [5]. An example is the refurbishment of components from decommissioned products for use in new ones. The aim of closed-loop PLM is to close information gaps between the phases and processes of the product lifecycle. This can be backwards, for example providing usage data to design processes, or forwards, for example providing production and assembly information to recycling processes. It deals with products as classes or variants, as well as individual product items ("item level").

1.2   Product Service System

The adoption of the service business by manufacturing companies is a common trend in many industrial sectors, especially those offering durable goods. This shift, referred to in literature as servitization process, is defined as "[…] the increased offering of fuller market packages or 'bundles' of customer focused combinations of goods, services, support, self-service and knowledge in order to add value to core product offerings" [6]. Servitization supports companies to strengthen their competitive position thanks to the financial, marketing and strategic benefits led by the integration of services in the companies' offer [6–9].

Differentiation against competitors, hindering competitors to offer similar product-service bundles and the increasing of customer loyalty are the main benefits of servitization. Today, more than ever, servitization is customer driven [10]. A research field that is often associated to the servitization process is the one related to the Product Service-Systems (PSS) [11]. The first definition of a PSS was given in 1999: "A product service-system is a system of products, services, networks of players and supporting infrastructure that continuously strives to be competitive, satisfy customer needs and have a lower environmental impact than traditional business models" [12].

Manzini points out that PSS is an innovation strategy that allows fulfilling specific customer needs [13]. Tukker observes that PSS is capable to enhance customer loyalty and build unique relationships since it follows customer needs better [14]. Another important contribution comes from Sakao and Shimomura that see PSS as a social system that enhances social and economic values for stakeholders [15].

The move towards the PSS entails an organizational change that makes a company shift from a product-oriented culture to a service-oriented one. The transition is quite a complex process that requires several changes and that usually happens in subsequent steps.

Martinez et al. identify the five categories of challenges a company has to deal with when moving along the servitization process, namely embedded product-service culture, delivery of integrated offering, internal processes and capabilities, strategic alignment and supplier relationships [16].

PSS often include value adding services based on ICT contributions, both in terms of enhanced information and knowledge generation/sharing, as well as of additional functionalities [17, 18]. PSS providers need to establish collaboration among specialized companies. In particular, Fisher et al. discussed approaches for service business development on a global scale. They take into account organizational elements, such as customer proximity or behavioral orientation [19].

The closer affiliation of customers and manufacturers/service providers offer potential to generate revenue throughout the entire lifecycle [18, 20]. Moreover, as stated by Baines et al., "… integrated product-service offerings are distinctive, long-lived, and easier to defend from competition based in lower cost economies …" [18]. The potential extension of the lifetime of tangible components of PSS, due to their integration with adding value services, opens interesting perspectives also about environmental sustainability improvements.

The advantages coming from PSS have been demonstrated in literature, yet for many companies efficiently managing the service operations is still a challenge. Best practices and empirical analysis are mainly carried out with a focus on larger companies. Nonetheless, the PSS topic is more and more recognized by SMEs that are looking for innovative business solutions to improve their competitive advantages.

1.3   Internet of Things for Smart Manufacturing

The term "Internet of Things" (IoT) was first used by the Massachusetts Institute of Technology in the year 1999. It was used in the sense of a networked system of autonomously interacting and self-organizing objects and processes, which was expected to lead to a convergence of physical things with the digital world of the Internet [21]. This extrapolates the idea of the Internet—a global, interconnected network of computers—to describe a network of interconnected things, such as everyday objects, products, and environments. At the heart of the concept lies the idea that objects—things—are capable of information processing, communication with each other and with their environment, and autonomous decision making. For instance, Intelligent Products are physical items, which may be transported, processed or used and comprise the ability to act in an intelligent manner. McFarlane et al. [22] define the Intelligent Product as:

> [...] a physical and information based representation of an item [...] which possesses a unique identification, is capable of communicating effectively with its environment, can retain or store data about itself, deploys a language to display its features, production requirements, etc., and is capable of participating in or making decisions relevant to its own destiny.

The degree of intelligence of a product may exhibit variations from simple data processing to complex pro-active behavior. Three dimensions of characterization of an Intelligent Product are suggested by Meyer et al. [23]: Level of Intelligence, Location of Intelligence and Aggregation Level of Intelligence. The first dimension describes whether the Intelligent Product exhibits information handling, problem notification or decisions making capabilities. The second shows whether the intelligence is built into the object, or whether it is located in the network. Finally, the aggregation level describes whether the item itself is intelligent or whether intelligence is aggregated at container level.

More recently Porter states that intelligence and connectivity enable an entirely new set of product functions and capabilities, which can be grouped into four areas: monitoring, control, optimization, and autonomy [24]. A product can potentially incorporate all four. Each capability is valuable in its own right and also sets the stage for the next level. For example, monitoring capabilities are the foundation for product control, optimization, and autonomy. A company must choose the set of capabilities that deliver its customer value and define its competitive positioning.

Smart, connected products have three core elements:

- *Physical* components comprise the product's mechanical and electrical parts. In a car, for example, these include the engine block, tires, and batteries.
- *Smart* components comprise the sensors, microprocessors, data storage, controls, software, and, typically, an embedded operating system and enhanced user interface. In a car, for example, smart components include the engine control unit, antilock braking system, rain-sensing windshields with automated wipers, and touch screen displays.
- *Connectivity* components comprise the ports, antennae, and protocols enabling wired or wireless connections with the product. Connectivity takes three forms, which can be present together:
  - One-to-one: an individual product connects to the user, the manufacturer, or another product through a port or other interface—for example, when a car is hooked up to a diagnostic machine.
  - One-to-many: a central system is continuously or intermittently connected to many products simultaneously. For example, many Tesla automobiles are connected to a single manufacturer system that monitors performance and accomplishes remote service and upgrades.
  - Many-to-many: multiple products connect to many other types of products and often also to external data sources. An array of types of farm equipment is connected to one another, and to geo-location data, to coordinate and optimize the farm system.

Connectivity serves a dual purpose. First, it allows information to be exchanged between the product and its operating environment, its maker, its users, and other products and systems. Second, connectivity enables some functions of the product to exist outside the physical device, in what is known as the product cloud.

Smart, connected products offer exponentially expanding opportunities for new functionality, far greater reliability, much higher product utilization, and capabilities.

These new types of products alter industry structure and the nature of competition, exposing companies to new competitive opportunities and threats. They are reshaping industry boundaries and creating entirely new industries. Smart, connected products have been shown to be applicable to various scenarios and business models. For instance, Kärkkäinen et al. describe the application of the concept to supply network information management problems [25]. Other examples are the application of the Smart Products to supply chain [26], manufacturing control [22, 27], and production, distribution, and warehouse management logistics [28].

Smart connected products are increasingly the focus of research into the collection of item-level product usage data for closed-loop PLM applications, servitization and product avatars [29, 30].

1.4 Life Cycle Cost (LCC) and Life Cycle Assessment (LCA)

Life Cycle Cost (LCC) analysis provides a framework for specifying the estimated total incremental cost of developing, producing, using and retiring a particular item. This methodology is useful to directly provide cost information to designers, in order to reduce the life cycle cost of the products they design [31].

There exist some difficulties in the application of LCC techniques to PSS, which usually includes the necessity of analyze various scenarios for effectively evaluating the impact of risks and uncertainties. These difficulties arise from some specificities of PSS, such as the modification of the role and responsibilities of customers and suppliers in the various PSS life cycle phases, the difficulty to foreseen the timing and overall frequency of use of some services, the lack of availability of life cycle data. The gap about LCC information among the various stakeholders during the PSS design phase can lead to unsatisfactory choices and prevent the full exploitation of PSS benefits [32, 33].

Life Cycle Assessment is "a process to evaluate the environmental burdens associated with a product system, or activity (process) by identifying and quantitatively describing the energy and materials used, and wastes released to the environment, and to assess the impacts of those energy and material uses and releases to the environment" (www.setac.org). To calculate impact ratios, LCA defines four phases that takes place iteratively: the goal/scope definition, the inventory definition and analysis, the impact assessment and the interpretation. Fundamental for the reliability and repeatability of calculating impact ratios is the completeness and quality of data and the transparency of processes and methodology applied.

Although LCA is a well-documented methodology (e.g., LCA handbook, 2010), repeatability is weakened because of the large freedom offered in choosing system boarders, parameter selection, data quantity and calculation methodology, which introduce uncertainties on estimated impact ratios and make difficult their comparisons. Moreover, due to the complexity and the diverse types of uncertainties inherent to LCA, simplifications and by analogy approaches are often required in order to use it [34]. This hinders the comparison of studies even when they address similar situations. The role of LCA in influencing design and more generally decision making towards a sustainability strategy is hindered by its current use, which often takes place as a posteriori side activity after product design fulfillment, as well as by the lack data models and tools able to capture and make transparent the choices and decision process during all the step of product lifecycle. These problems are exacerbated while considering PSS due to some specific challenges, such as:

- Wide difference of PSS typologies implying modifications to the required activities and the involved actors [35];
- Strong influence of the context of application of PSS for determining the encounters and the methodologies to be followed [36],
- Unsatisfactory integration of sustainability issues in current PSS design methodologies.

1.5 The Manutelligence Project

The Manutelligence Project aims at supporting enterprises to design and to develop suitable Product-Service Systems, addressing customers' needs and stakeholders' requirements. Manutelligence aims to integrate best in class methodologies and tools from research and industry, resulting in a secure, collaborative Product/Service Design and Manufacturing Engineering Platform.

The Manutelligence consortium consists of a group of highly qualified industrial and academic research organizations that has been specifically affiliated to meet the challenges of the project.

All the involved RTD partners have a strong experience in publicly funded projects, both at a European and a national level, with high innovative and application capabilities. The RTD partners have the core competences and expertises required to cover the knowledge domains of this project (information and communication technologies, product lifecycle management, product & service innovation management, data and knowledge management, etc.). The application partners are concentrated on the industry-driven implementation and evaluation, to prove the resulting research concepts.

The partners are divided as follow:

- 4 Research partners broken down as follows:

  - 2 Universities: Politecnico di Milano and Supsi.
  - 2 Research Institutes: VTT and BIBA.

- 3 ICT Industrial partners: Dassault Systèmes, Holonix and Balance.
- 5 Industrial Companies: Ferrari, Mayer Turku, Lindbäcks Bygg, Fundacio Privada Centre CIM, Rina Consulting.

Concerning the geographical distribution of the consortium partners, Manutelligence gathers partners form seven different countries: Finland, France, Germany, Italy, Spain, Sweden, and Switzerland.

1.5.1 Manutelligence Research Objectives

The main research topics addressed during the projects have been:

- Improve efficiency and develop new methodology for the PSS design process, with a specific focus on the integration of IoT technologies (Chap. 2).
- Achieve a complete integration of Product Lifecycle Management and Service Lifecycle Management, developing concepts, methodologies and tools to support PSS development (Chap. 2).
- Adapt and integrate existing design, data analysis and life cycle assessment tools to realize closed-loop PLM for PSS (Chap. 2).
- Enable designers and engineers access data from the traditional enterprise IT systems, but also from the IoT enabled systems. The objective is to manage all data,

information and knowledge related to the P-S and its lifecycle in manufacturing. (Chap. 3).

- Extract feedback from P-S customers, analyzing data coming from IoT systems, in order to speed up the design of P-S, and to decrease the costs and to better understand customer needs (Chap. 3).
- Extend and improve the use of Manufacturing and Service Execution Simulation and optimize it through comparisons with test bench and real usage data (Chaps. 3 and 4).
- Measure and simulate costs and sustainability issues, through Life Cycle Cost (LCC) and Life Cycle Assessment (LCA), collecting data from both traditional sources and smart connected products. The combined use of various tools allows effectively sharing LCC and LCA information to all the stakeholders in a simple way, supporting their decision making processes (Chap. 5).

1.5.2 The Manutelligence IT Platform

To achieve the described objectives, Manutelligence aims to integrate best-in-class methodology and tools from research and industry, resulting in a secure, collaborative Manufacturing Engineering Platform. This platform enables designers and engineers to access data from both the traditional enterprise IT systems (CAD, CAX, PLM, MES, etc.) and from smart, connected products. In Table 1.1, the architecture of the Manutelligence platform is presented.

The platform consists of the integration of different tools components, which will be exhaustively described in Chap. 4.

The core technical achievements of Manutelligence are:

- Inclusion of tools for the process design and manufacturing execution. These tools are intrinsically integrated with the PSS design phase and can leverage on the IoT information coming from the operations.
- Access information through a 3D interface representing the digital representation of the product, containing both information from the digital product model stored in the PLM and those coming from Intelligent Products (IoT technologies).

Table 1.1 Manutelligence's tools integration

Partner tool name	Brief description of component	Provided by partner
3DEXPERIENCE	Managing the Product Service Design and Manufacturing processes	Dassault Systemes
I-Like	Managing the Internet Of Things (IoT) data gathering and elaboration	Holonix
MaGA	Managing the environmental impact analysis	SUPSI
LCPA	Managing the Product Service life cycle cost analysis	BALANCE

- Support the interaction between the engineering and the environmental (LCA) or business (LCC) analysts, as well as to provide tools and methods to enable iterative calculation and optimization of these aspects. The platform results a suitable tool to collect, share data and information helping analysts to retrieve data and to define boundaries of the analysis.
- Features of the platform can be applied in many different industrial cases, improving the manufacturing efficiency and quality, addressing the needs captured from the products usage by the end users.

References

1. Terzi S, Bouras A, Dutta D, Garetti M, Kiritsis D (2010) Product lifecycle management—from its history to its new role. Int J Prod Lifecycle Manag 4(4):360–389. ISSN (Online): 1743-5129, ISSN (Print)
2. Rossi M, Riboldi M, Cerri D, Terzi S, Garetti M (2013) Product lifecycle management adoption versus lifecycle orientation: evidences from Italian companies. In: Proceedings of product lifecycle management for society 10th IFIP WG 5.1 international conference, PLM 2013, Nantes, France, July 6–10, 2013, pp 346–355
3. Wellsandt S, Nabati E, Wuest T, Hribernik KA, Thoben KD (2016) A survey of product lifecycle models: towards complex products and service offers. Int J Prod Lifecycle Manag 9(4):353
4. Hong-Bae J, Kiritsis D, Xirouchakis P (2007) Research issues on closed-loop PLM. Comput Ind 58(8–9):855–868
5. Pokharel S, Mutha A (2009) Perspectives in reverse logistics: a review. Resour Conserv Recycl 53:8
6. Vandermerwe S, Rada J (1988) Servitization of business: adding value by adding services. Eur Manag J 6(4):314–324
7. Oliva R, Kallenberg R (2003) Managing the transition from products to services. Int J Serv Ind Manag 14(2):160–172
8. Gebauer H, Friendli T, Fleisch E (2006) Success factors for achieving high service revenues in manufacturing companies. Benchmarking: Int J 13(3):374–386
9. Malleret V (2006) Value creation through service offers. Eur Manag J 24(1):106–116
10. Lewis M, Howard M (2009) Beyond products and services: shifting value generation in the automotive supply chain. Int J Automot Technol Manage 9(1):4–17
11. Tukker A, Tischner U (2006) New business for Old Europe: product-service development, competitiveness and sustainability. Greenleaf Publishing, Sheffield
12. Goedkoop M, van Haler C, te Riele H, Rommers P (1999) Product service-systems, ecological and economic basics, pre consultants, The Hague. Report for Dutch Ministries of Environment (VROM) and Economic Affairs (EZ)
13. Manzini E, Vezzoli C (2003) A strategic design approach to develop sustainable product service systems: examples taken from the 'environmentally friendly innovation' Italian prize. J Clean Prod 11(8):851–857
14. Tukker A (2004) Eight types of product–service system: eight ways to sustainability? Experiences from SusProNet. Bus Strategy Environ 13(4):246–260
15. Sakao T, Shimomura Y (2007) Service Engineering: a novel engineering discipline for producers to increase value combining service and product. J Clean Prod 15(6): 590–604. https://doi.org/10.1016/j.lclepro.2006.05.015
16. Martinez V, Bastl M, Kingston J, Evans S (2010) Challenges in transforming manufacturing organizations into product-service providers. J Manufact Technol Manag 21(4):449–469

17. Jansson K, Kalliokoski P, Heimilä J (2003) Extended products in one-of-a-kind product delivery and service networks. In: Building of the Knowledge Economy. Proceedings of the eChallenges conference in Bologna, IOS Press, Amsterdam

18. Jannsson K, Thoben K-D (2005) The extended products paradigm, an introduction. In: Arai E, Kimura F, Goossenaerts J, Shirase K (eds) Knowledge and skill chains in engineering and manufacturing, IFIP International Federation for Information Processing, vol 168. Springer, Berlin, Heidelberg, pp 39–47

19. Fisher T, Gebauer H, Fleish E (2012) Service business development, strategies for value creation in manufacturing firms. Cambridge University Press, Cambridge

20. Baines TS, Lightfoot H, Steve E, Neely A, Greenough R, Peppard J, Roy R, Shehab E, Braganza A, Tiwari A, Alcock J, Angus J, Bastl M, Cousens A, Irving P, Johnson M, Kingston J, Lockett H, Martinez V, Michele P, Tranfield D, Walton I, Wilson H (2007) State-of-the-art in product-service systems. Proc Inst Mech Eng, Part B: J Eng Manuf 221(10):1543–1552

21. Brand L, Hülser T, Grimm, V, Zweck A (2009) Internet der Dinge – Perspektiven für die Logisitk. Zukünftige Technologien Consulting

22. McFarlane D, Sarma S, Chirn JL, Wong CY, Ashton K (2003) Auto ID systems and intelligent manufacturing control. Eng Appl Artif Intell 16:365–376

23. Meyer GG, Främling K, Holmström J (2009) Intelligent products: a survey. Comput Ind 60:137–148

24. Porter ME, Heppelmann JE (2014) How smart, connected products are transforming competition. Harvard Business Review 92(11):64–88

25. Kärkkäinen M, Holmström J, Främling K, Artto K (2003) Intelligent products—a step towards a more effective project delivery chain. Comput Ind 50:141–151

26. Ventä O (2007) Intelligent and Systems. Technology theme—final report. VTT, p 304. VTT Publications, Espoo

27. Hribernik K, Cassina J, Røstad CC, Thoben K-D, Taisch M (2012) Potentials of item-level PLM and servitization in the leisure boat sector. In: Proceedings of the 1st international through-life engineering services conference. Shrivenham, UK, 28

28. Wuest T, Hribernik K, Thoben K.-D (2012) Can a product have a Facebook? A new perspective on product avatars in product lifecycle management. In: IFIP WG 5.1 international conference, PLM 2012, Montreal, QC, Canada, July 9–11, 2012, pp 400-410

29. Wuest T, Irgens C, Thoben K.-D (2013) An approach to quality monitoring in manufacturing using supervised machine learning on product state data. J Intel Manufact, online first. https://doi.org/10.1007/s10845-013-0761-y

30. Asiedu Y, Gu P (1998) Product life cycle cost analysis: state of the art review. Int J Prod Res 36(4):883–908

31. Erkoyuncu JA, Roy R, Shehab E, Wardle P (2009) Uncertainty challenges in service cost estimation for product service systems in the aerospace and defence industries. In: Proceedings of the 1st CIRP Industrial Product-Service Systems (IPS2) conference, Cranfield University, 1–2 April 2009, pp 200

32. Schuh G, Boos W, Kozielski S (2009) Life cycle cost-orientated service models for tool and die companies. In: Proceedings of the 1st CIRP Industrial Product-Service Systems (IPS2) conference, Cranfield University, 1–2 April 2009, pp 249

33. Sahni S, Boustani A, Gutowski T, Graves S (2010) Textile remanufacturing and energy Savings. Environmentally Benign Laboratory. Laboratory for Manufacturing and Productivity, Sloan School of Management, MITEI-1-g-2010

34. Fischbach M, Puschmann T, Alt R (2013) Service lifecycle management. Bus Inf Syst Eng 5(1):45–49

35. Mukhtar M, Ismail MN, Yahya Y (2012) A hierarchical classification of co-creation models and techniques to aid in product or service design. Comput Ind 63(4):289–297

36. Vasantha GVA, Rajkumar R, Lelah A, Brissaud D (2012) A review of product-service systems design methodologies. J Eng Des 23(9):635–659

Chapter 2
Engineering and Business Requirements Definition, Analysis and Validation

Iris Karvonen, Tapani Ryynänen, Heidi Korhonen, Matteo Cocco and Donatella Corti

Abstract The objective of Manutelligence platform is to manage manufacturing intelligence; all data, information and knowledge related to the Product Service (PS) and its lifecycle. The platform is based on two existing platforms and some analysis tools (for example LCA and LCC). It was developed according to the needs of four use cases in different industrial fields (automotive, ship, smart house, 3D-printing). The chapter describes the four-phase methodology to define the common aggregated requirements for the platform. The phases include requirement elicitation, structuration and organization, analysis and refinement and validation. In the elicitation phase the requirements were identified from the use cases, in the structuration and refinement phases they were further consolidated, categorized and processed towards aggregated requirements and in the validation phase the resulting aggregated requirements were compared to the original use case requirements. The chapter also shows the main results of each phase.

I. Karvonen (✉) · T. Ryynänen · H. Korhonen
VTT Technical Research Centre of Finland Ltd., P.O. Box 1000, 02044 Espoo, Finland
e-mail: Iris.Karvonen@vtt.fi

T. Ryynänen
e-mail: Tapani.Ryynanen@vtt.fi

H. Korhonen
e-mail: Heidi.Korhonen@vtt.fi

M. Cocco
Dassault Systèmes, Milano, Italy
e-mail: Matteo.Cocco@3ds.com

D. Corti
SUPSI—University of Applied Sciences and Arts of Southern Switzerland, Via Cantonale, 2C, 6928 Manno, Switzerland
e-mail: Donatella.Corti@supsi.ch

© The Author(s) 2019
L. Cattaneo and S. Terzi (eds.), *Models, Methods and Tools for Product Service Design*, PoliMI SpringerBriefs, https://doi.org/10.1007/978-3-319-95849-1_2

2.1 Introduction

This chapter describes the methodology and main results of the definition and processing of engineering and business requirements for Manutelligence platform. The
chapter is focused on the phase before platform implementation; thus also the validation here is about validation of final requirements against the use case scenarios
and requirements. The platform validation against the requirements is not discussed
here.

As a starting point for development, Manutelligence had two existing platforms
and some existing analysis tools. These have been consolidated, complemented and
adapted to become the Manutelligence platform. Thus the Manutelligence approach
was different from the basic software requirements definition, which often starts
from the scratch (new application or module) or has the description or original user
requirements of the existing platform available.

The development has been guided by the needs of participating industrial pilots.
The Manutelligence project included four industrial pilots from different industrial
fields (automotive, ship, smart house, 3D-printing). These cases were the sources for
industrial requirements in the project. All the pilots already use various engineering
tools in the product design. The idea was not to collect all the potential functions that
an engineering platform could cover, but to identify new needs with relation to their
current tools and practices. Thus the collected requirements from the use cases do
not compose a complete set of requirements for a generic PS engineering platform.

The requirement engineering process was carried out using a common methodology through the following phases: **requirement elicitation, structuring and organization, refinement and validation**. The intermediate requirement processing and
consolidation phases were needed because the different pilot scenarios were focused
on different processes and industries, with various stakeholders and user needs, which
resulted in a heterogeneous set of elicited requirements, difficult to use as such for
the platform development. In the process attention was given to keep the traceability.

In the requirement elicitation phase, the idea was to identify requirements with a
wide scope, not restricting in what could be implemented in the current project. On
one hand the wider scope gave more input for the platform development, on the other
hand the pilots were in the elicitation phase not yet able to make the decision about
what will be implemented in Manutelligence. Thus it was clear from the beginning
that not all wishes in original requirements are implemented in this project with the
restricted project resources. Thus the requirements should not be considered as static
and final but more as an iterative and evolutionary set of needs.

2.2 Challenges

The main objective in requirement elicitation was to receive requirements that arise
from the real needs of end users and the focus was not on the formal quality. The end

users were not specialists in requirement engineering, but in PS design and engineering methods and tools. Requirement identification is often challenging, as the end users are not able to express their needs directly. Instead they need to be dragged out using different methods, taking into account the end-user business objectives. Thus user friendly methods were needed. The approach generated a set of heterogeneous requirements, which required further analysis and processing.

The sources of heterogeneity came already from different concepts and terminologies used in different sectors, but also from different groups of stakeholders, PS systems and different engineering processes and practices. Additionally, the pilot companies represent different company sizes and have differences in their preparedness for the utilization of information technology. The different groups also produced requirements with different levels of detail.

Given the above, the four datasets received were challenging to structure and consolidate. Therefore the structuration and analysis of the requirements required manual and iterative processing of data. As the structuration and analysis phases were mainly performed by researchers using different methods, the end users were again in the main role in the validation phase to check that the consolidated requirements were sufficient compared to the original pilot scenarios.

2.3 Methodology

2.3.1 Four Phase Approach

The Manutelligence approach was to integrate and adapt existing technologies to fulfill the development needs of the four pilot cases in selected PS engineering process parts. The approach affected the Manutelligence methodology for requirements engineering.

The selected approach was to apply a four-phase methodology with the phases: Elicitation, Structuration, Analysis and Refinement and Validation. The objectives of the four Requirements Engineering phases were:

Elicitation. During this process heterogeneous needs and opportunities coming from different stakeholders involved in the PS development were identified from the pilots.
Structuration. The main objective of the structuration was to unify and integrate the information collected in the previous step from disparate sources and organize them into a common structure that can be used for analysis.
Refinement and Analysis. The target of this activity was to refine and verify the previously elicited requirements. The refinement consists of the assessment of the completeness, coherence and feasibility of the stakeholders' requirements and their prioritization according to different criteria.
Requirements validation. The purpose of the validation was to ensure that the structured and consolidated requirements were sufficient for the end users (pilots) and could fulfill the defined scenarios. Thus this phase was the validation of the

consolidated requirements against the pilot needs (scenarios and stories), not the validation of the implementation. Later in the project a validation of the platform against the consolidated requirements was performed. This platform validation is out of the scope of this chapter.

2.3.2 Requirements Elicitation Techniques in Manutelligence

The task of requirements elicitation is the identification of requirements' sources and the elicitation of requirements according to the identified stakeholders and other requirements sources [3]. The elicitation can be performed using different methodologies such as interview, questionnaire, observation, brainstorming, prototyping, mind-map and checklist. In this phase, human activity is fundamental and it is necessary to identify users involved in the process and establish a relation between them and the developers [2].

The elicitation was started with a **pre-elicitation** phase to identify the context in which the Manutelligence project will be developed. In the pre-elicitation, information was collected using a short questionnaire about the understanding of the holistic PS and what are the stakeholder expectations from the project. It was a kind of "close interview" technique in which the stakeholders answered to a predefined set of open-ended questions (3 questions).

After this preparation phase the actual **elicitation** was carried out. The following elicitation techniques were used: questionnaire, process mapping, pilot stories and pilot scenarios.

In Manutelligence a comprehensive questionnaire with about 30 questions was used to investigate the industrial practices in product and service (PS) development and data management in the four pilots. The questionnaire included the following parts:

Part 1. Design process at glance.
Part 2. Managing knowledge in a design and development context.
Part 3. Managing the development of the PS.
Part 4. Evaluating the lifecycle of the PS.

In the process mapping activity, the PS lifecycle of each pilot was modeled to understand the main life cycle phases and their interaction and the focus of process developments needed. Because of the different levels of complexity in the pilot cases the resulting models varied in the level of detail.

The pilot story is a customer and user centric methodology, useful to understand the whole domain of the project. A pilot story basically is a storytelling with a description of how the user would interact with the Manutelligence platform rather than how it works internally or how it is designed. Telling the story, the end user is able to present the desired future operations. Going through the story, it was possible to identify requirements enabling the story to come true.

In parallel with the requirement identification, pilot scenarios for the Manutelligence project were described. The scenarios described the candidate as-is and to-be use cases to be implemented in the project pilots, offering information in a more structured format: purpose and objectives, actors involved, systems etc. This information was also used in the elicitation of the requirements.

2.3.3 Structuration Methods

The objective of the structuration phase was to organize the requirements coming from different sources into a common structure and to consolidate them to a moderate number of requirements. Thus firstly the structure had to be defined, then all the requirements were allocated to the structure and finally they were aggregated. In the beginning, each requirement was given a unique identifier that also connects it to the original pilot. This identifier followed the requirement throughout the process so that the original requirement could always be traced back.

In the structuration two approaches were integrated: top-down and bottom-up. In the **top-down** approach, the concepts and structures given by the project were identified. These could be found for example in the interviews or questionnaires. The structures were compared to find similarities, which did not have to be exactly the same but on the same dimension, like for example different process phases of product-service lifecycles.

In the **bottom-up** approach, the structures emerging from the data were identified. The task utilized an adaptation of the Thematic analysis method [1]. An understanding of the data (original pilot requirements) was required in the task, often leading to necessity to familiarize oneself with the pilot stories and scenarios.

The bottom-up approach thus meant analyzing the unstructured requirements to identify similarities, categories and structures. The goal was to form a generic structure or hierarchy of categories that suits for all the use cases and supports the development of the Manutelligence platform.

In the next phase the information available from both the given structures (top-down) and from the list of unstructured requirements (bottom-up) was analyzed and relations, similarities and differences were identified. The final structure was formed, based on understanding the knowledge from both approaches and the complete data.

The pilot requirements were organized to the defined structure. The organization also tested if the structure was sufficient, if it was possible to put each requirement somewhere in the structure.

Finally the original pilot requirements belonging to the same subcategory were **aggregated**. The aggregated requirements are not as detailed as the original ones but they aim to integrate similar needs from different pilots. The links to original requirements were maintained.

2.3.4   Analysis and Prioritization Method

The objective of the third phase was to further refine and prioritize the structured requirements coming from the previous phase. The aim of the prioritization was not to remove any requirements but to create an overall view of their high level importance. The final decision of the requirements to be implemented during the project was taken along the pilot development.

The requirements were first reviewed in order to make the level of detail more homogeneous and to eliminate potential duplications. Next a trade-off analysis was performed to identify on one hand mutually supportive and on the other hand conflicting requirements. In the trade-off analysis each couple of requirements was considered and the corresponding relationship was qualitatively evaluated. The correlation was analyzed considering the mutual impact of requirements during the development of the platform. A positive correlation means that the parallel fulfillment of the two requirements is mutually supportive and vice-versa. Values ranging from -2 to $+2$ were used.

For the prioritization two types of criteria were defined: (1) Manutelligence-related criteria and (2) Pilot-related criteria. Manutelligence-related criteria come from understanding the general objectives of the project. Aggregated requirements were used in this phase. Pilot-related criteria are based on the needs of the pilot cases; thus the original unstructured requirements were used here. These requirements were considered on how much they can positively impact on the design process of the PS in the pilot.

Findings coming from the two prioritization analyses were finally merged to form the final rank. A bonus system was used that favors more those requirements that are addressed as important by both the Manutelligence-related and the pilot-related criteria. This final rank achieved provided evidence about what are the most relevant requirements to be fulfilled within the Manutelligence project since it summarized all the previous analyses based on different points of view.

2.3.5   Requirements Validation Method

Validation has different roles over the application development process. In Manutelligence the first validation took place in the requirement definition phase and it was about **validation of requirements**, not software. Thus the objective of the requirements validation here was not to check that the Manutelligence platform and related tools fulfill the requirements, but that the aggregated requirements fulfill the end user needs. Also the prioritization defined in the previous task was checked. This was needed, as the composition, structuring, aggregation and analysis (including prioritization) of requirements from different use cases were performed by the supporting partners, not the use case owners themselves.

Different methods for the validation were applied. First an individual review using a walk-through approach was used to check the sufficiency of aggregated requirements (not the priorities). The review was performed by a group of researchers representing the partners supporting the end users in Manutelligence. In the review each of the use cases was handled separately. Two pieces of source material for each case were used: (1) the pilot stories and (2) the end user scenario descriptions (to-be). The approach in the review was first to walk through the pilot story step by step and to identify the main functionality needed for each step. Thereafter the list of needed functionalities was compared to the list of aggregated requirements to see if there is a requirement available, which enables taking the step. After that, the same was done for the end user scenarios (to-be). As the aggregated requirements are on a higher level, telling more about "what" than "how", the idea was to find a high level requirement, which could cover the lower level functionality.

It is clear that not all aggregated requirements were needed for each use case, but the other way around; at least one requirement was needed for each step. Otherwise a shortage was recorded.

To include the end users (pilots) in the validation, a specific validation workshop was organized. The workshop contained the following three main sessions:

- Presentation of the aggregated requirements,
- Industrial partners crosschecking the Use Case requirements,
- Industrial partners checking the prioritization of the requirements.

The main task was the crosschecking of the aggregated requirements by the industrial partners. The participants were divided into sub-groups, one for each pilot and one for software developers, five groups in all. The methodology used was a form of Requirements Walk-Through and Reading Technique. The participants were asked to review the partner specific Pilot Stories and Use Case scenario descriptions (to-be) to check the sufficiency of the aggregated requirements. The groups were equipped with printouts, in A3 size, of Pilot Stories, Use Case scenarios and the list of aggregated requirements. Figure 2.1 depicts the methodology.

The participants read through their Pilot Story and Use Case scenarios, section by section. For each encountered step or function in the text, the participant checked that a corresponding requirement could be found in the list of aggregated requirements. These were marked with a circled 1, 2 and 3 etc. as seen in Fig. 2.2. If an aggregated requirement covering the issue could not be found, then a note was made. The number of how many times an aggregated requirement was referenced to was counted for each Industrial partner.

The third and final step in the workshop for each Industrial partner was to point out the most important aggregated requirements. Each industrial partner was asked to mark the five top important aggregated requirements for its specific use cases. The given rankings were summarized into an overall ranking.

Fig. 2.1 Reading technique used in the workshop

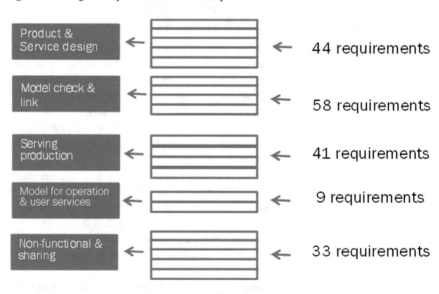

Fig. 2.2 Requirement categories and number of unstructured requirements in each category

2.4 Results from the Definition of Business and Engineering Requirements

2.4.1 Results from Requirements Elicitation

The requirement elicitation generated more than 200 requirements coming from the four industrial pilots (automotive 23, ship 129, smart house 25, 3D-printing 18) and from LCA/LCC technical workpackages (9). The number of requirements coming from one use case (ship) was much higher than from other use cases. This was due to using a requirement hierarchy and more detailed low level requirements. As expected, the requirements were quite heterogeneous and focusing on different process parts in the PS engineering.

2.4.2 Results from Structuration and Organization

As described earlier, the structuring and categorization of requirements were performed by reconciling the results of top-down and bottom-up approaches.

For the top-down approach, concepts coming from the project were studied. Manutelligence project is focused on Product-Service design using manufacturing intelligence and through the development of a platform to support the whole Product Service lifecycle. Thus, from Manutelligence context the following main concepts could be identified:

- Product service (PS) (answering to question WHAT).
- PS Lifecycle (WHEN).
- PS actors/stakeholders (BY WHOM).
- PS related knowledge/information/data.
- Platform (HOW; this is for what the requirements are).

Based on these, from top-down there were several alternatives for requirement categorization, for example based on Product Service type, information type, stakeholders etc. Product Service type of categorization would not support the integration of requirements of different use cases. Classification according to the stakeholders would be difficult as in most cases the objective is to support the information sharing, communication and collaboration between different stakeholders in all tasks. The division according to information type cannot be strict as many of the requirements consider different kinds of information. Especially there is a need to be able to handle and link them together.

Thus, it seemed that the most suitable candidate for the top-down structure, which was significant for all the use cases, is based on the lifecycle phases.

In the bottom-up approach the requirements coming from different sources were analyzed to identify a structure, which could suit for all the use cases and assist in the aggregation of their requirements. Additionally it should be understandable.

Table 2.1 The top 12 words and terms in the unstructured requirements

Word	Occurrence	Theme	Occurrence
Design	91	3D-viewing	106
Model	70	Feedback management	104
Feedback	53	Production planning and control	66
Data	50	Platform content	55
Product	47	Product Configuration, BOM	49
Customer	44	Link to drawings	36
Production	32	Customer interaction	33
View	30	Validation and Verification, inspection	23
3D	30	Design	23
Access	26	Training, Guidance	21
Management	22	LCA/LCC, Environmental issues	14
Link	19	Change management	11

To identify the important topics bottom-up, thematic analysis was applied. Two practical methods were used:

- Calculation of specific relevant words from the requirement collection to identify subjects that have high interest. About 50 words were searched.
- Definition of a group of terms/themes based on the requirements (more than one word) and analyzing their occurrence. 17 terms were searched.

Table 2.1 shows the top 12 occurred words and themes. The number of occurrences for each word is affected by the heterogeneity of the requirements. This is mainly because the different use cases have given their requirements on different levels of detail, using more or less words. Thus the requirements including longer and more detailed expressions have more impact on this analysis.

The analysis identified as frequent some words that could be expected to be present in many requirements, like design, model, data product/production and view and access. However, there were also words with high frequency that were not as expected, like feedback, customer, and link. On the other hand, some terms like service and lifecycle did not belong to the top group.

On the right side of Table 2.1 again the top 12 occurrences of themes (not exact words) are presented. The themes seem to be in line with the identified words.

Structuring based on Product Service life cycle seemed to be most suitable in the top-down approach. The main intrinsic grouping in the use cases also followed the life cycle approach. In the thematic analysis some of the words and themes identified were clearly related to one specific lifecycle phase and some were related to more than one phase. The thematic analysis also revealed different types of functions, especially related to the design phase. Design phase is not only of engineering/design but also preparing and using the designed product information/model for intelligent actions in different life cycle phases. Thus, when selecting the high level structure for the integrated requirements, it was seen useful to divide the design phase functions to two groups; those related to the real design phase (creation of the product design/model) and those using the product model for linking additional information/documents, to perform analysis and checking and to prepare for life cycle services.

The defined five high level requirement categories are described below. The origin of the included requirements is shown with the codes: A-Automotive, S-Ship, C-Construction/Smart House, F-3D printing, LCA.

Product and service design into model
This includes all the requirements related to Product development/design/engineering/building the 3D model for the product, for example: product requirement management (F), product configuration (F+C), design based on construction method (S), creation and conversion of design for 3D printers (F), design changes and version management (A, C).

Model checking and linking
This includes checking and analyzing the product using the 3D model and linking information/feedback to it, for example: tests, analysis, simulation, data into the platform (A), cost calculation (C), LCA/LCC-analysis (LCA), sustainability analysis (F), customer feedback through 3D and gaming experience (S).

Serving production through model
This includes using the 3D-model to support manufacturing/construction/installation, for example: installation support (S), inspection support (S), production feedback (S), project planning and management (C), developing the production cycle (F).

Model for operation and user services
This includes all the requirements for the operation and usage phase, like measurement with sensors and IoT (C), operation feedback (S), monitoring (S) which use the product model and related information.

Sharing and non-functional requirements
This includes all non-functional requirements, also related to sharing and access to information, like access to the product model and needed information through it (S), all information embedded and managed in the platform (A), sharing Fablab-models (F) and security aspects.

Categories 1–3 belong to the Beginning of Life and 4 to Middle of Life. We identified no requirements for the End of Life.

After defining the high level structure, it was validated by organizing all the single requirements to this common structure. Mainly it was easy to place the requirements to the structure but in some cases a requirement was set in two different groups. There were mainly two reasons for this: 1. Some requirements included in fact more than one requirement in the same sentence. 2. Some requirements were not completely clear and interpretation was needed.

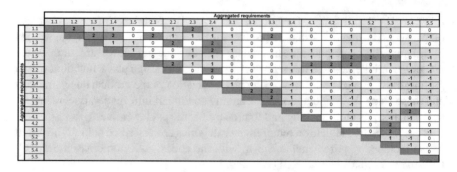

Fig. 2.3 Trade-off among macro-requirements. The scale used is the following: 2: strong positive correlation; 1: positive correlation; 0: no correlation; −1: negative correlation

As a result of the aggregation all the ~200 requirements were aggregated into $5 + 4 + 4 + 2 + 5 = 20$ requirements (Fig. 2.2).

2.4.3 Results from Prioritization

The trade-off analysis (correlation between the requirements) was performed by the research partners. The results are shown in Fig. 2.3. Each aggregated requirement has been given a number (for example 1.3), where the first number represents the requirement category.

The prioritization criteria were defined as follows: (1) Manutelligence-related prioritization criteria including Implementation time, Implementation cost, Technical gap, Usability in other sectors, Scalability, PSS fitting, and Enabling collaboration. Weights for the criteria were also defined. (2) Pilot related criteria, for example, Design time, Change management agility, Improved communication with customers and Improved communication among designers. Each pilot defined its own set of weights for the criteria, since it was expected that the relative importance of the performance associated to the criteria vary depending on the specific context.

The Manutelligence-related criteria were applied for the 20 aggregated requirements while the pilot-related for the original pilot requirements separately for each pilot. The individual scores where then summarized to the aggregated requirements.

For the final ranking additionally a bonus system was defined. The basic idea was that the score of an aggregated requirement obtained from the Manutelligence-related criteria consideration was increased by a bonus if the same requirement was considered to be important also by one or more pilots.

It should be noted that the final list and the relative rank was not frozen at this point: the overall requirements engineering process followed a spiral approach and during the development new interests also came up. Also, no requirements were eliminated from the list in this phase, even though some of them were assigned a lower score. The highest scores were received by the following requirements:

4.2 The service provider shall be able to manage the services and their data on the platform.

2.1 Designers and experts shall be able to perform and manage real time cost calculation/LCC and LCA assessment along the design using the platform and the data from design and previous projects.

1.1 The designer shall be able to systematically manage product requirements and trace design changes and versions within the platform.

1.5 The collaboration network/community of designers shall be able to support and contribute to the design on the platform.

2.5 Validation Results

In the individual walk through the pilot stories and to-be scenarios, 1–7 steps were identified for which there was no clear corresponding requirement. Mostly these needs were quite detailed or very specific. In addition there were some steps for which link to an existing requirement could be identified, but the requirement should somehow be extended to cover a specific aspect. There were two main types of comments:

- The requirement should include "service" in addition to product. This is very important as Manutelligence aims to support Product-Service design.
- In addition to other stakeholders, also customer or end user should have access to the PS information. This is relevant as the customer interaction and participation is even more important when providing PS than when providing products.

Based on these comments, the aggregated requirements were reformulated to cover also services and customers/users.

All the four use cases participated in the validation workshop. First the 20 aggregated requirements were discussed and clarified with their source: which use cases and original requirements have affected to each requirement. Most of the requirements were considered understandable but also some comments regarding them were received from the use cases. These mainly included terminology.

The validation of requirements was performed in five groups: one for each use case + one group for platform providers. The use case representatives were supported by the research partners. The methodology described before was followed: the use cases went through the pilot story and identified the aggregated requirements, which supported the pilot story activity. The links between the pilot story steps and the requirements were marked with the same number. The idea was also to identify missing requirements for supporting pilot steps.

The number of identified links for each aggregated requirement from the use cases was identified. The following was observed:

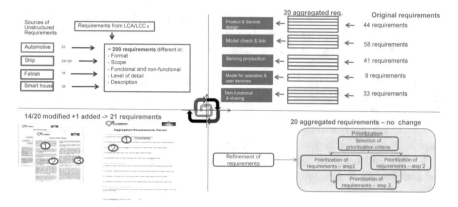

Fig. 2.4 Requirement identification, processing and validation process

- The total number of occurrences for a single requirement was between 1 and 14.
- No requirement was found unnecessary.
- 4 requirements were relevant only for one use case.
- There was only one requirement that was needed for all use cases.
- 6 requirements had a link to all but one (3) use case.

The validation revealed some missing requirements and comments for the requirement text.

Finally, the results (comments and missing requirements) of the different validation steps were collected together and the requirements were reviewed to study how they should be changed based on the comments. As a whole, 14 requirements were reformulated and 1 new requirement was added.

Furthermore, to validate the prioritization, the use cases were asked to select the 5 most important requirements from their viewpoint. The selections of use cases were quite scattered. There was no requirement that belonged to the 5 most important of all the use cases. 6 requirements were selected by 2 use cases and 6 were selected by no use case. Thus in each pilot case the decision was made what to implement in Manutelligence project, and what to leave for later.

2.6 Conclusion

A summary of the four phases of the business and engineering requirements is presented in Fig. 2.4 (starting from up-left). The process went through requirement elicitation, structuration and organization, analysis and prioritization and finally validation. In the process a large number of diverse and heterogeneous requirements were identified, organized and aggregated into a manageable set of requirements, well-structured and prioritized PS requirements.

The final list of 21 aggregated and validated requirements is the following:

1. **Product service design into model**

   1.1 The user shall be able to systematically manage product and service requirements and trace design changes and versions within the platform.

   1.2 The user shall be able to manage the product and service structure to create product configuration and BOM on the platform.

   1.3 The designer shall be able to use knowledge in the design based on previous models and the platform shall provide automatically the rules and design methods.

   1.4 The platform shall be able to make easily data conversions (CAD files, product models, visualization models, manufacturing models).

   1.5 The collaboration network/community of designers shall be able to support and contribute the design on the platform.

2. **Model analysis and linking**

   2.1 Designers, customers and other users shall be able to perform, manage and view real time LCC and LCA. Here the data from design and previous projects should be available.

   2.2 The platform shall use the product model to link, manage and allow access to all the results of (quality) tests and simulations.

   2.3 The customer shall be able to view the visual product model (including virtual walk/driving) and give feedback on it using the platform.

   2.4 The designer and the production shall be able to link and manage information, data and documents in the product model supported by platform specifications and rules.

3. **Serving production through model**

   3.1 The production personnel and the customer shall be able to use the product model on the platform to support and monitor production, installation and to give feedback for the design.

   3.2 Project manager and customer shall be able to use the model on the platform to plan, monitor and manage the contract and the production.

   3.3 The manufacturer/production coordinator/user shall be able to manage the production resources and suppliers through the platform.

   3.4 The production/quality management shall be able to manage and follow the quality and failure data on the platform.

4. **Model for operation and user services**

   4.1 The user/ service provider shall be able to monitor the behaviour of the product using the product model and linked sensors with access to the platform.

   4.2 The service provider and the customer shall be able to manage the services and their data on the platform.

5. **Non-functional requirements (including sharing information)**

5.1 All information shall be managed and embedded in a common platform, which is applicable for different industrial sectors.

5.2 The platform shall provide user management and allow access to stakeholders according their rights and needs.

5.3 The platform shall support sharing and communication, also remotely/online/off-line.

5.4 The platform shall manage data security and quality (including metadata).

5.5 User interface of the platform services shall be easy to use and include user support, like tutorials.

5.6 The platform should allow to display PSS-related advertisements for the customer and support the selling process for additional products and services in dependency of a PSS's life cycle phase and its actual condition (for example by visualizing them in the product context).

These aggregated requirements were used in the development phase to support the platform and pilot development. In the final phase of the project the implemented solution (pilots and the platform) was validated against the aggregated requirements.

References

1. Aronson J (1995) A pragmatic view of thematic analysis. Qual Rep 2(1):1–3. Retrieved from http://nsuworks.nova.edu/tqr/vol2/iss1/3. Accessed 13 Dec 2017
2. Sawyer P, Kotonya G (2001) Software requirements. SWEBOK, 9
3. Sommerville I, Kotonya G (1998) Requirements engineering: processes and techniques. Wiley, New York

Chapter 3
Life Cycle Management for Product-Service Systems

Stefan Wellsandt, Laura Cattaneo, Daniele Cerri, Sergio Terzi, Donatella Corti, Christian Norden and Reinhard Ahlers

Abstract Product-Service Systems (PSS) and the Internet of Things (IoT) are two related concepts. This chapter describes an approach to manage PSS along its life cycle. It includes a design methodology for PSS and a systems modelling method. The former supports designers in defining PSSs that incorporate monitoring, control, optimization or autonomy. It includes a new method to assess a product's functionality in terms of the data needed for its realization. The latter adopts life cycle thinking and employs a modelling language to outline the PSS and its various components and actors. A life cycle performance analysis could benefit from the model by extracting cost information from it for further analysis. This chapter highlights challenges related to PSS life cycle management observed during the Manutelligence project. They concern the design methodology and the applied life cycle modelling method.

S. Wellsandt (✉)
BIBA—Bremer Institut für Produktion und Logistik GmbH at the University of Bremen, Hochschulring 20, 28359 Bremen, Germany
e-mail: wel@biba.uni-bremen.de

L. Cattaneo · D. Cerri · S. Terzi
Department of Economics, Management and Industrial Engineering, Politecnico di Milano, 20156 Milan, Italy
e-mail: laura1.cattaneo@polimi.it

D. Cerri
e-mail: daniele.cerri@polimi.it

S. Terzi
e-mail: sergio.terzi@polimi.it

D. Corti
Department of Innovative Technologies SUPSI, University of Applied Sciences and Arts of Southern Switzerland, 6928 Manno, Switzerland
e-mail: donatella.corti@supsi.ch

C. Norden · R. Ahlers
Balance Technology Consulting GmbH, 28203 Bremen, Germany
e-mail: christian.norden@bal.eu

R. Ahlers
e-mail: reinhard.ahlers@bal.eu

© The Author(s) 2019
L. Cattaneo and S. Terzi (eds.), *Models, Methods and Tools for Product Service Design*, PoliMI SpringerBriefs, https://doi.org/10.1007/978-3-319-95849-1_3

3.1 Product-Service Systems and the Internet of Things

A PSS is an integrated product and service offering that delivers value in use to the customer. There are several potential benefits of PSSs. First, there is an increase of revenues, because services tend to have higher profit margins and can provide a stable source of revenues. Second, they are means to differentiate offers in mass-markets, which are typically characterized by commodities and technologies. Third, services bring a decrease of variability and volatility of cash flow throughout the life of a product resulting in a higher shareholder value [1]. PSS is a complex concept, because it is composed of several parts that need to be managed: a product and one or more related services, a network of stakeholders and a supporting infrastructure [2]. The different product aspects, such as the concept and design phases and the life cycle, have been analyzed deeper by many authors [3], while the service topic is more recent and first attempts exist for structuring the service design [4].

The IoT technologies promote the business model of PSS. They enable new services related to the connectivity and the usage of environmental data. Connected smart products offer expanding opportunities for new functionality, greater reliability, higher product utilization and capabilities that cut across and transcend traditional product boundaries [5]. Terms like Cyber-Physical Systems (CPS), IoT and Virtual Reality have become ordinary in industries. IoT products raise a new set of strategic choices related to

- how value is created and captured,
- how the amount of data they generate is utilized and managed, and
- what role companies should play as industry boundaries are expanded [5].

Smart products have three main components: physical components, smart components and connectivity components. Smart components are directly connected with services related to the physical parts, while connectivity allows exchanging information between the product and its environment, and enables services to exist outside the physical product itself [5].

Intelligence and connectivity enable an entirely new set of product functions and capabilities, which can be grouped into four areas: monitoring, control, optimization, and autonomy [5]. A product can potentially incorporate all four. Each capability is valuable in its own right and sets the stage for the next level. For example, monitoring capabilities are the foundation for product control, optimization, and autonomy.

- **Monitoring**: smart, connected products enable the monitoring of a product's condition, operation and external environment through sensors and external data sources. A product can alert users or others stakeholders to changes in circumstances or performance. Monitoring also allows companies and customers to track a product's operating characteristics and history. It improves the understanding of how the product is actually used. The collected data has implications for: (1) design by reducing over-engineering and improving market segmentation through the analysis of usage patterns by customer type; (2) after-sale service by allowing the dispatch of the right technician with the right part to improve the first-time fix rate. Monitoring data may also reveal warranty compliance issues, as well as new sales opportunities, such as the need for additional product capacity because of high utilization.

- **Control**: smart, connected products can be controlled through remote commands or algorithms that are built into the device or reside in the product cloud. Algorithms are rules that direct the product to respond to specific changes in its condition or environment (for example, "if pressure gets too high, shut off the valve" or "when traffic in a parking garage reaches a certain level, turn the overhead lighting on or off"). Control through software embedded in the product or the cloud allows the customization of product performance to a degree that was not cost effective or often even possible before. The same technology also enables users to control and personalize their interaction with the product in many new ways. For example, users can adjust their smart light bulbs via smartphone, turning them on and off, programming them to blink red if an intruder is detected, or dimming them slowly at night.

- **Optimization**: smart, connected products can apply algorithms and analytics to in-use or historical data to improve output, utilization, and efficiency. In wind turbines, for instance, a local microcontroller can adjust each blade on every revolution to capture maximum wind energy. Each turbine can be adjusted to improve its performance and to minimize its impact on the efficiency of those nearby. The rich flow of monitoring data from smart, connected products, coupled with the capacity to control product operation, allows companies to optimize product performance in numerous ways, many of which have not been previously possible.

- **Autonomy**: monitoring, control and optimization capabilities combine to allow smart, connected products to achieve a previously unattainable level of autonomy. At the simplest level an autonomous product does operation using sensors and software on real time. More-sophisticated products are able to learn about their environment, self-diagnose their own service needs, and adapt to users' preferences. Autonomy not only can reduce the need for operators, but it can improve safety in dangerous environments and facilitate operation in remote locations. Autonomous products can also act in coordination with other products and systems. The value of these capabilities can grow exponentially as more and more products become connected.

Connected products require the design of a new technology infrastructure made of multiple layers. It is necessary to identify sensors, software, data storage, user interfaces as well as ports, protocols and kind of connections to design the PSS.

The IoT could underpin the PSS design by using information feedback at any stage of the Product-Service life cycle. Moreover, IoT provides information for continuous improvement, closer relationship with stakeholders, resource efficiency and the ability to meet sustainability. IoT provides the opportunity to obtain information about some relevant parts of the system in real time, during all the life cycle processes, thus representing a differential in order to build, design, implement and improve a PSS.

3.2 Life Cycle Thinking, Modelling and Management

The following paragraphs adopt the findings of a literature survey published in the International Journal for Product Life cycle Management [6]. The study investigated the constituents of life cycle models.

Life cycle is a concept used in Biology where it describes the recurring change of states for certain organisms (see [7] for biological life cycle). In engineering, the organisms are exchanged for tangible goods (products). Processes, such as design, production, use, repair, recycling and disposal substitute the organism's state changes. From the perspective of business economics, the concept of the product life cycles was introduced by Theodore Levitt in the 1960s [8]. Levitt's approach describes the life cycle of products by four phases: introduction into the market, rapid growth of sales volume, market saturation and market decline. The introduction of Levitt's product life cycle model supports the strategic management of products. Manutelligence focused on the engineering life cycle.

Life cycle thinking means that people have a life cycle model in mind that affects the scope of their activities during, for instance, system development and production. In Manutelligence, a **life cycle model** describes activities related to a PSS in a simplified way. It consists of elements that represent the transformation of the product, software or service over time, whereas the notion of 'transformation' is adopted from operations management [9]. A *transformation* changes inputs (e.g. raw material) into outputs (e.g. machine) of a higher quality and a higher value. Processes and/or the stakeholders performing specific processes represent the transformation.

Technological advances, such as the IoT, resulted in developments that influence life cycle thinking. Effects of these developments and common characteristics of a PSS's life cycle are summarized in Table 3.1. The presented concepts are not meant to be comprehensive.

Examples for the influence of life cycle thinking on engineering are Life Cycle Assessment and Life Cycle Costing. The adoption of life cycle thinking in industry and academia led to the emergence of a plethora of different life cycle models as investigated by Wellsandt et al. [6]. Manutelligence focused on the flow of codified information along the life cycle.

Life cycle management, in the Manutelligence project, focuses on the management of information along the PSS life cycle. The specialization on PSS is a rather new discipline, because the adoption of Servitization, Intelligent Products and the Internet of Things is not very high in most industries. Consequently, the experience with this PSS life cycle management is low, compared to the information management for tangible products or software. One of the challenges of PSS life cycle management is the fact that a PSS consists of integrated hardware, software and service components. This is especially true, if the PSS has connectivity-based features. In this context, integration has several meanings, for example, the:

1 Integration of ICT into traditional products (Intelligent/Smart Products).
2 Integration of hard- and software development (Concurrent Engineering).
3 Integration of PSS software into third party software (Interoperability).
4 Integration of services into existing business processes.

Table 3.1 Concepts of PSS life cycles

Name	Description
Processes and stakeholders	The cornerstones of the life cycle model are processes and the stakeholders being affected by them
Flows and stocks	Information, material and energy are the constituents that are exchanged between processes to create value and quality
Process abstraction	The life cycle model consists of different process abstraction levels. In Manutelligence, three layers of abstraction were differentiated. These are life cycle phase, process, and activity
System abstraction	Complex systems consist of sub-systems. They can be represented on different abstraction layers in a life cycle model. A common system abstraction is the differentiation of product, assembly and component
Information abstraction	A system class describes systems with shared characteristics. It represents a high level of abstraction. An instance describes an entity that features a unique characteristic, such as a product identifier. It is the lowest abstraction layer for products used in Manutelligence
Product states	A product's state is described by using item-level information, collected at a specific time and place [10]. Each state consists of a set of relevant "state characteristics" measured at specific checkpoints during the life cycle
Structure	Life cycle processes can be structured as a linear or circular sequence. If two linear life cycles share a process, a cross structure is established
Nesting	A system's life cycle can be described as a nested structure of sub-system life cycles. This way, a complex structure of several related life cycle models is created

Each of these integrations represents a complex problem with many variables that can relate to each other. The problems are dynamic because they change with the rapid advances in ICT technology (e.g. the fast spread of Blockchain technology). Solving the integration problems benefits from an extensive understanding of, for instance, the PSS's components and stakeholders. For this reason, one assumption of Manutelligence was that life cycle modelling helps developers and other stakeholders to understand the complexity of a PSS during the design phase. The actual modelling task is a supplement to standard procedures in product design. These procedures include, for instance, the collection and analysis of requirements, and the development of system components and their interfaces.

The Manutelligence partners developed an approach to support the design of IoT-related PSSs. It consists of a PSS design methodology and the concurrent life cycle modelling. Figure 3.1 illustrates this approach.

The following sub-chapters describe the proposed design approach. It starts with the new design method for the IoT Assessment. This chapter omits a description of the Business Model Canvas (BMC) and Quality Function Deployment (QFD) methods, because literature has plenty of descriptions for them already [11, 12]. The second part of this chapter introduces the life cycle modelling and the life cycle performance analysis approach.

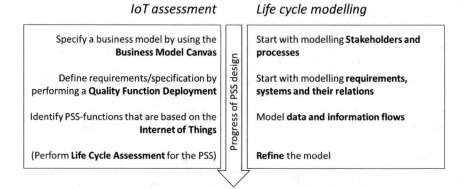

Fig. 3.1 Manutelligence PSS design approach

3.3 Approach to Design IoT-Related PSS

3.3.1 Design Methodology

This part of the chapter is based on the Manutelligence public Deliverable 2.2 [13]. The main goal of the methodology is to structure the PSS's concept and design processes with a focus on the identification of the IoT technologies needed to catch and elaborate data for customer need satisfaction, and for the design of suitable services. The idea is to connect different activities and methods, to elicit data and information. The methodology is modular, different activities are optional, and they do not need to be performed in a specific order. Within these main activities, we include the definition of the business model, the definition of the stakeholders' requirements and the definition of the IoT capabilities. Besides these three tasks, there are other tasks to be considered, especially concerning the design of the actual solution components. The main activities should be completed and integrated with other important information, as well as the addition of hard and software functionalities and business processes.

We will focus on the definition of the IoT capabilities, which represents the innovative part of the methodology developed during the Manutelligence project.

3.3.2 IoT Assessment Support

This activity represents the core part of the proposed methodology, since it defines the bases for the PSS design phase. The outputs of this process will be the Bill of Material (BOM) of the product, the definition of the connected services and the identification of data, information and knowledge that should be managed and analyzed during the Product-Service life cycle. Connect products require the design of a new technology infrastructure made up of multiple layers. It is necessary to identify sensors, software, data storage, user interfaces as well as ports, protocols and kind of connections to design the PSS. Furthermore, it is necessary to satisfy the stakeholders' requirements, which should be previously analyzed, for instance with a QFD.

Usually the Product Design and the Service Design processes are separately performed within specialized teams. This produces solutions that often are not really integrated. The innovation of the Manutelligence methodology aims to improve the integration between Product and Service Design, in order to give to customers the best PSS solution.

Starting from the requirements defined in a previous phase, the team will implement the Product-Service Design with the primary aim of defining the IoT capabilities that should be integrate in the product life cycle, in order to properly design the connected services.

The idea is to develop a graphical tool that can be used by designers to identify and classify the IoT capabilities that should be embedded in the product. This tool has to include:

- The classification of the capabilities of the smart product, which can be grouped into four areas [5]: monitoring, control, optimization and autonomy. This classification specifies the kind of service a company wants to implement thanks to the IoT technology.
- The definition of data destination. It is necessary to define the stakeholder who needs a specific kind of information (e.g. customer, designer, and maintainer). This is important also to identify the format with which data or analysis results should be shown and the support that must be used, in order to read and share data and results. For example, customers would like to have access to data directly from their smartphone.
- The identification of the product parts in which sensors should be installed in order to collect data.

To implement the graphical tool, we exploit the BOM, which should be developed in the Product Design phase. The BOM is particularly useful to identify the product's parts. Parts and components define a specific number of details and levels. For each of these levels, we design a specific graphic structure. For instance, for a smart car, at level 0 we have the vehicle. Level 1 composes of the Power Unit, the Gear Box and the Car Body (Fig. 3.2).

For each level, the product's parts are reported as input for the designed table. The first column identifies the data destination and the stakeholders involved in the service. The entries of the table are colored following the capabilities classification (i.e. monitoring is green colored, control is yellow colored, optimization is orange colored and autonomy is red colored,) and the entries are filled indicating the variable. If a box remains white, this means that no IoT capabilities are defined for that entry. The IoT capability is selected, in order to satisfy the product and service requirements. This classification is an important initial stage, since a specific IoT capability defines the correspondent technology to select, for example, sensors or data storage systems and clarify the kind of connection that must be developed.

		# Level				
		Part 1	Part 2	Part 3	Part 4	Part 5
Data destination	Software	Control		Monitoring		
	Designers					
	Maintenance				Monitoring	
	App		Monitoring		Optimization	

Fig. 3.2 Illustration of the IoT graphical tool

3.4 Life Cycle Modelling Support

This chapter is based on a journal paper published in the International Journal for Product Life Cycle Management [6] and the Manutelligence public Deliverable 2.1. The paper's goal was to investigate life cycle models and identify common elements and structures of these models. For this purpose, 71 models from literature were investigated leading to the aforementioned concepts of PSS life cycle models.

A modelling language that supports many of the concepts described in Table 3.2 was used for the actual modelling task of the use cases. For this purpose, the Lifecycle Modelling Language (LML) was selected in Manutelligence [14]. It is a derivative of the Systems Modelling Language (SysML). LML introduces an Ontology that describes the system life cycle with specified entities and their relations amongst each other. The modelling task was supported by the browser-based systems engineering tool Innoslate (https://www.innoslate.com/). It has native support for LML—the software provider appears to be involved in the development of the LML standard. LML's features and the capabilities of the modelling tool Innoslate were mapped against the aforementioned life cycle concepts. A summary of this mapping is presented in Table 3.2.

LML suggests 12 entities and several visualization techniques for system life cycle modelling. The Innoslate-functions used in Manutelligence mainly base on the entities defined in LML. Instances of the entities build the life cycle model. Some Innoslate-functions relate to visual techniques used by the software (e.g. abstraction layers). The mapping between life cycle concepts and LML is mainly achieved by using existing entities and their relations from the LML Ontology. Process and system abstraction is achieved through the relations between activities and assets (e.g. decomposes and decomposed of). Different structures of a life cycle can be created by linking activities through an activity diagram. No support was identified in LML for the nesting concept. Innoslate allows users to name a process as "life

Table 3.2 Mapping between life cycle concepts, LML and Innoslate

Life cycle concept	LML features	Innoslate functions
Processes and stakeholders	Activity and Asset	Via LML
Flows and stocks	Input/Outputs and Activities	Via LML
Process abstraction	Relation between Activities	Visual abstraction layers
System abstraction	Relation between Assets	Visual abstraction layers
Information abstraction	No support identified	No support identified
Product states	No support identified	No support identified
Structure	Activity diagram	Via LML
Nesting	No support identified	Visual separation of processes

cycle" and decompose it. In this case, several life cycle containers can be created with related Activity entities. A shortcoming of LML and Innoslate is the lacking support for item level information. This also limits the support for the product state concept. LML and Innoslate can be replaced by other modelling approaches that support the concepts presented in Table 3.1.

3.4.1 Life Cycle Analysis Support

This chapter is based on a conference paper for the International Conference on Engineering, Technology and Innovation held 2017 [15]. The paper's goal was to present and discuss an approach for life cycle analysis using a shared life cycle model. Life cycle analysis for traditional products includes methods, such as "Life Cycle Costing" (LCC) and "Life Cycle Impact Assessment" (LCA). These methods were selected for Manutelligence, because they represent widely acknowledged life cycle analyses.

Life Cycle Costing (LCC) is an accounting method, which considers every cost flow throughout the life cycle of a product [16]. **Life Cycle Performance Assessment** (LCPA) enhances the LCC-approach by extending the assessment to cash inflows. The key performance indicator of the LCPA-approach is the net-present value. It reflects the point of time of each considered cash flow throughout the life cycle. Consequently, the cash flows of the assessed object discount with an interest rate depending on its specific point in time. Moreover, LCPA pursues an approach that allows to compare investments with each other.

The quality of LCPA results depends on the input data quantity and quality. Every cash flow throughout the life cycle has to be identified with its cash flow type and its point in time in the life cycle. Usually, the information needs to be tied to a specific life cycle phase and quantified with its time-based value. In many cases, this requirement leads to a time series of cash flows, which needs to be archived in the ex-post perspective or forecasted in the ex-ante perspective. In the next steps, the gathered data has to be translated into a software-tool to perform the LCPA.

The general LCPA approach can be adapted for the assessment of PSS. Besides the relevant input data for products, such as investments, energy costs, maintenance and other operating costs, the analysis of the service part requires input data that is more focused on, for instance, personnel costs, equipment usage to perform the service, travel costs, as well as service fees as revenues. As a result, the comparative LCPA approach enables the evaluation of different PSS concepts with each other.

The main reference to carry out a **Life Cycle Impact Assessment** (LCA) is the ISO 14000 family of standards. They provide a set of international guidelines quite stable, subjected to periodic updates, internationally recognized and used as reference at global level. The identified and renowned structure of an LCA study is organized into four phases:

- **Definition of the goal and scope of LCA**. It defines why the LCA is performed and what the system boundaries are.
- **Life Cycle Inventory (LCI) analysis**. The exchanged natural elements and their quantities of the resources and emissions entering (e.g. raw material, energy and ancillary material) and leaving the system of interest (e.g. emission, waste, products and co-products) have to be defined. This step is complex and time consuming, since it involves data collection from several different actors and related processes along the supply network. The allocation process of shared resources is a critical point and has an impact on the final result.
- **Life Cycle Impact Assessment (LCIA)**. It is meant to calculate the impacts on the environment generated by the identified LCI data.
- **Interpretation**. In the final phase, the report with the quantified impacts is prepared and the critical review of the LCA results is performed.

The ISO 14050 standard has been developed with a physical product focus, even though the provided definition of "product" states "any good or service". This means that services are conceptually considered, yet the PSS concept is not considered explicitly. For this reason, the ISO-based LCA approach might not be applicable to the PSS context directly. The main difficulty when carrying out an LCA for a PSS is how to integrate the service component into the LCI. Corti et al. proposed an approach to support the LCI phase for PSSs [17]. It could formalize the integration of information related to the service part of the offer. Categories of information and their positioning along the life cycle of the PSS are taken into consideration showing the backbone structure that is recommended to carry out the LCI for a PSS. The life cycle divides into its three main phases, namely Beginning of Life (BOL), Middle of Life (MOL) and End of Life (EOL), whilst the data categories are related to either the product component or the service component.

Guidelines for gathering the information to support the LCA methodology are still missing [18, 19]. Dal Lago et al. propose a conceptual work in this direction [20]. The authors develop a set of guidelines to gather the information needed to instantiate and support an LCA analysis from a PSS life cycle perspective.

Table 3.3 Approach for model-based life cycle analysis for PSS

Step	Realized
Identify elements needed for analysis	Manual task
Create life cycle model	Modelling tool
Export model into xml file	Modelling tool
Import model from xml file	Analysis tool
Run analysis	Analysis tool

A shared life cycle model supports the PSS life cycle analyses in Manutelligence. Table 3.3 summarizes the proposed approach.

The proposed approach is most efficient, if multiple analyses refer to the same life cycle model. In this case, redundant information collection and modelling is avoided. The *first* step of a model-based life cycle analysis for PSS is the identification of relevant elements from the PSS life cycle. Relevance depends on the necessity to represent an element in the analyses. For instance, it might be necessary to differentiate a product and its components, stakeholders, and activities during the usage phase. In some cases, it could be relevant to omit activities from the PSS life cycle, because their expected influence on cost or environmental impact is low in the targeted application case. The *second* step is the creation of a life cycle model that contains information about the analysis elements. Manutelligence realized this through a model grounded on the LML Ontology. It was developed and maintained in Innoslate. The tool features an xml-file export of the entire model and the Ontology—this effectively represents the *third* step of the proposed approach. Once the model is in a common data format, it can be imported into the actual analysis tool. The import of the life cycle model represents the *fourth* step. Finally, the analysis is initialized with the information stored in the imported life cycle model. This *fifth* step is the final one of the proposed approach for model-based life cycle analysis of PSS. The approach was tested with an LCPA tool.

3.5 Challenges for Product-Service Systems Management

This chapter partially bases on a conference paper for the International Conference on Engineering, Technology and Innovation held 2017 [15]. The following paragraphs cover different challenges concerning PSS life cycle management. The first parts summarize issues encountered during the evaluation of IoT components of a PSS. Challenges related to creation of life cycle models follow it. Each of the paragraphs aims to provide suggestions how the encountered challenges could be addressed in the future.

3.5.1 *IoT Assessment*

The IoT assessment method is the original contribution of our methodology, since other phases are actually well-known methodologies already used in several different fields [11, 12]. The IoT assessment graphical tool should be able to design how information and data collected through IoT technologies have to be integrated along the PSS life cycle. We present a list of questions that the IoT assessment tool should be able to answer and we identify which points need further developments and improvements.

Why? *Why we want to use IoT technologies? Which is the service we want to develop? Which particular function we want to perform?* Even if these points should come straightforward from the HoQ requirements, the IoT assessment should highlight these goals, in order to keep focus on them. The graphical tool registers the IoT capabilities that should be implemented in the PSS, as suggested in [5].

Who? *Who is involved, which are the different stakeholders that take part in this process?* The actual graphical tool, during its application to Manutelligence use cases, turned out to be a little bit confusing, since we have put as data destination designers, maintenance, applications and software. Actually, the presence of hardware and software is always mandatory, because data are collected and analyzed with software support. Only after these first steps, processed data could be visualized/used by someone else, for example, designers could use analytics results to improve product design and product development.

What? *What we want to measure? Which data and information we want to collect?* Looking at the graphical support, we have no information regarding which variables and data should be collected along the process.

Where? *Where we have to insert and install sensors? Which components of the product are involved in the collection of information?* We answer to this question using information coming from the BOM. At this moment we investigate all the different BOM levels, but how it turned out from its application to Manutelligence use cases, this could request too much effort than what is really needed. For example, the level of details in a car's BOM could be deep and even unnecessary for this aim. In principle, the BOM could be used only to identify and take note of parts involved in the data collection process.

When? *Which phase of the product life cycle is involved? When information is collected? When information and data could be used?* It could be necessary to highlight, which are the PSS life cycle phases involved in the process, also to easily identify stakeholders and different flows of data and information between them. For example, data collected from machine-embedded sensors (MOL phase) could be sent to maintainers to realize predictive maintenance (MOL phase). Data collected during the product usage (MOL phase) could be sent to designers, in order to improve product design and product development (BOL phase).

With What? *What kind of technology should be developed? What about hardware and software installed and used to collect, analyze and share data? Which data analysis should be performed?* In the graphical tool, we do not mention the technologies that should be put in place, in order to collect data and allow connectivity (hardware, ports, protocols and kind of connections) even if this point is important, in order to check if all the different devices are able to be connected one with each other. As some of the use cases have highlighted during project development, it is also important to perform a feasibility analysis, to compare benefits and costs of the technology investment. The graphical tool should also contain an initial definition of the statistical techniques that should be used to analyze data. When machine learning techniques are implemented from some software, data are directly processed from an algorithm and results are made available for further analysis (for example, after Cluster Analysis it could be decided to perform a Classification Rules Analysis). In some cases, the presence of a data scientist could be requested, which could take some raw or processed data and perform further analysis or post processing operations, in order to organize the output such that it is easily understandable. Only after these intermediary steps, data could be available to be read and used by other stakeholders.

3.5.2 Life Cycle Modelling

Model complexity. Complexity refers to the number of elements and relations in a life cycle model. The investigated use cases in Manutelligence feature around 100 elements and more than 100 relations (using the LML Ontology). The required number of elements depends on the model's purpose and on the PSS. A support of multiple life cycle analyses (and other tasks) may lead to the need to include additional elements and relations. Higher complexity will likely make the management of the model more difficult (e.g. adding, removing and updating elements). Adding elements to the life cycle model can be avoided if two or more tasks share an element. Additional relations, however, might be necessary in the case of element sharing.

Model dynamics. The concept of dynamics refers to the need that a model's elements and relations may evolve over time (depending on the model's purpose). In the course of the PSS design and operations, for instance, the model is subject to extension or reduction. Elements and relations change depending on the needs of the decision makers involved in the related tasks. This is especially the case when the model is also used as a PSS planning tool—as experienced in Manutelligence. Maintaining a life cycle model, i.e. keeping it updated to the most recent planning state, can be a time consuming task. The time needed to maintain a model depends, among others, on the complexity of the model. For instance, it is a difficult task to add an element and its relations to a model with several hundreds of elements and even more relations. In the case of Manutelligence, the maintenance of the model was not experienced as difficult. Possibly, because the model is not yet complex enough to affect maintenance.

Modelling tools. Creation and maintenance of a life cycle model can be difficult tasks that should be supported with software tools. The decision to use LML for our study limits the number of tools to support this task. The only tool that integrates it "out of the box" is called Innoslate. Key persons from the company behind this tool are contributors to the LML specification. The logic of LML may be replicated with other tools, such as Enterprise Architect and the ARIS toolbox. Life cycle modelling tools should, in general, satisfy the needs of the stakeholders involved in the PSS life cycle. In the case of life cycle analysis, the stakeholders (analysts) need to access (read) life cycle model information that is relevant for their analysis task. A complete model export with subsequent file parsing, as realized in Manutelligence, is only a temporary solution. This is, amongst other reasons, because the parsing could miss an element that is not following the convention of the parsing mechanism. A simple example is a typo in the element name. In order to solve this and other data quality problems, a reliable life cycle model exchange mechanism is needed.

References

1. Cavalieri S, Pezzotta G (2012) Product–service systems engineering: state of the art and research challenges. Comput Ind 63(4):278–288
2. Goedkoop MJ (1999) Product service systems, ecological and economic basics. Ministry of Housing, Spatial Planning and the Environment, Communications Directorate
3. Ulrich KT, Eppinger SD (2008) Product design and development, 4th edn, Internat edn. Boston: McGraw-Hill/Irwin
4. Aurich JC, Mannweiler C, Schweitzer E (2010) How to design and offer services successfully. CIRP J Manuf Sci Technol 2(3):136–143
5. Porter ME, Heppelmann JE (2014) How smart, connected products are transforming competition. Harvard Bus Rev 92(11):64–88
6. Wellsandt S, Nabati E, Wuest T, Hribernik KA, Thoben KD (2016) A survey of product lifecycle models: towards complex products and service offers. Int J Prod Lifecycle Manage 9(4):353
7. Bell G, Koufopanou V (1991) The architecture of the life cycle in small organisms. Philos Trans R Soc Lond B: Biol Sci 332(1262):81–89
8. Levitt T (1965) Exploit the product life cycle. Harvard Bus Rev 43:81–94
9. Grabner T (2014) Operations Management: Auftragserfüllung bei Sach- und Dienstleistungen, 2 edn (Aktualisierte Aufl.). Wiesbaden: Springer Gabler
10. Wuest T, Irgens C, Thoben K-D (2014) An approach to monitoring quality in manufacturing using supervised machine learning on product state data. J Intell Manuf 25(5):1167–1180
11. Osterwalder A, Pigneur Y (2010) Business model generation: a handbook for visionaries, game changers, and challengers. Wiley, New York
12. Akao Y (2004) Quality function deployment. Productivity Press
13. Manutelligence consortium, "Manutelligence project" (2015) [Online] Available: http://www.manutelligence.eu/. Accessed 21 Jan 2016
14. Lifecyclemodeling.org, "Lifecycle Modeling Language" (2017) [Online] Available: http://www.lifecyclemodeling.org/. Accessed 29 Aug 2017
15. Wellsandt S et al (2017) Model-supported lifecycle analysis: an approach for product-service systems. In: International conference on engineering, technology and innovation, Madeira
16. International Organization for Standardization (2008) ISO 15686-5:2008—buildings and constructed assets—service-life planning—part 5: life-cycle costing. International Organization for Standardization

17. Corti D, Fontana A, Montorsi F (2016) Reference data architecture for PSSs life cycle inventory. Procedia CIRP 47:300–304
18. Kjaer LL, Pagoropoulos A, Schmidt JH, McAloone TC (2016) Challenges when evaluating product/service-systems through life cycle assessment. J Cleaner Prod 120:95–104
19. Dal Lago M, Corti D, Wellsandt S (2017) "Reinterpreting the LCA standard procedure for PSS. Procedia CIRP

Chapter 4
A Platform for Product-Service Design and Manufacturing Intelligence

Maurizio Petrucciani, Lorenzo Marangi, Massimiliano Agosta and Marco Stevanella

Abstract Manutelligence is designed for the Product-Service business, allowing enterprises to develop innovative Product-Services, more sustainable, addressing customer needs. This platform will enable designers and engineers to access through natural 3DEXPERIENCE to data from both the "traditional" enterprise IT systems (CAD, CAX, PLM, MES, etc.) and IoT enabled systems for physical products information and knowledge management during its whole life cycle phases. The activities carried out by Manutelligence will improve the product and service development by connecting them together through cross disciplinary feedback loops by means of modular collaborative secure ICT manufacturing intelligence. The four software pillars of the Manutelligence platform are the 3DEXPERIENCE, by Dassault Systèmes, I-Like, by Holonix, MaGA, by SUPSI, LCPA, by Balance.

4.1 P-S Collaborative Design and Manufacturing Platform

Manutelligence aims at supporting the emerging trend of the Product-Service business, allowing enterprises to develop **innovative and more sustainable Product-Services, addressing customer needs**. Some of these services can be provided only after timely and accurate analysis of customers' product usage in order to acquire useful information for new product improvements or services provision. Often the misalignment between product and service development processes and unability of concurrent engineering between both processes arise due to the lack of information exchange among the product and service life cycle phases. This generates longer time to market for the product and service, misalignment between the product and service life cycle phases, lack of sharing knowledge and product and services not adapted to the business environment/customers' needs. Manutelligence aims to integrate best in class methodologies and tools from research and industry, resulting in a secure, cross disciplinary collaborative Product-Service Design and Manufacturing Engineering

M. Petrucciani (✉) · L. Marangi · M. Agosta · M. Stevanella
Dassault Systemes Italia Srl, Via dell'Innovazione, 3, 20126 Milan, Italy
e-mail: Maurizio.Petrucciani@3ds.com

© The Author(s) 2019   45
L. Cattaneo and S. Terzi (eds.), *Models, Methods and Tools for Product Service Design*, PoliMI SpringerBriefs, https://doi.org/10.1007/978-3-319-95849-1_4

Platform. This platform should enable designers and engineers to access through **natural 3D experiences** to data from both the "**traditional**" **enterprise IT systems** (CAD, CAX, PLM, MES, etc.) and **IoT enabled systems** for physical products information and knowledge management during its whole life cycle phases. The activities carried out by **Manutelligence** will **improve the product and service** development by **connecting them together** through cross disciplinary feedback loops **by** means of **modular collaborative secure ICT manufacturing intelligence**.

Therefore the innovative point is reached providing a lifecycle transversal infrastructure, able to provide to the different involved stakeholders (designers, engineers, manufacturing managers, testing, maintenance users and service team) a coherent, secure and content driven access to information (Fig. 4.1).

Moreover Life Cycle Cost (LCC) and Life Cycle Analysis (LCA) are usually a long and difficult process and cannot be directly used during the design phase by engineers, but needs specialized analysts due to the difficulty of retrieving data and defining the boundaries of the analysis. Objective of Manutelligence is to support the interaction between the engineering and the environmental (LCA) or business (LCC) analysts as well as to provide tools and methods to enable iterative calculation and optimization of these aspects.

4.2   Product Service Solution for Industrial Scenarios

The Manutelligence Platform has been developed based on industrial use cases and scenarii, with the target to be applicable in the related industrial sectors.

The Manutelligence Platform design was organized to support the industrial cases of the Manutelligence project provided by the partners *Ferrari*, *Meyer*, *FabLab (Fundacio CIM)* and *Lindbäcks*. Chapter 6, about Use cases applications, is showing the use case details.

The *Ferrari* case has been developed using the CAD design and simulation capabilities for the frequency and modal analysis as well as the IoT functionalities to capture and elaborate the driver usage of the car, recording information via telemetry. The *Meyer* case has been managed implementing the enterprise change management process; this process can be triggered directly via the IoT, automatically creating the issue object containing the data coming from problems captured during the ship operations. The *FabLab* made usage of the CAD design and BOM management

Fig. 4.1 Lifecycle phases

functionalities to develop the lamp and the MOVEO (robot arm) 3d-printing case, predicting the environmental impact with sustainability tools and measuring the in operations energy consumption via the IoT solution. The *Lindbäcks* case leveraged the IoT functionalities to monitor in real time an apartment usage via a remote device, like a smartphone.

The CAD capabilities either for design or for visualization are used in *Ferrari*, *Meyer* and *FabLab* cases. The simulation capabilities are used in the *Ferrari* case for the frequency and modal analysis.

The Manutelligence Platform includes tools for the process design and manufacturing execution. These tools are intrinsically integrated with the design phase and can leverage on the IoT information coming from the operations.

The features of the platform can be applied in many different industrial cases, improving the manufacturing efficiency and quality, addressing the needs captured from the products usage by the end users.

Based on the result obtained, it appears that the Manutelligence exploitation can be extended not only to the companies of the same industrial sector, meaning automotive, shipbuilding, construction, but also to other industries where the service component can be a source of new business, like, for example, the white goods sector. In fact anywhere there is the need to gather information about the usage of the products to improve design and manufacturing there is an opportunity to adopt a solution like Manutelligence platform.

4.3 P-S Collaborative Design and Manufacturing Platform Components

The Manutelligence Platform is the result from the integration of following Partner's tools (Table 4.1):

Table 4.1 Different tools composing the Manutelligence Platform

Partner tool name	Brief description of component	Provided by partner
3DEXPERIENCE	Managing the product service design and manufacturing processes	Dassault Systemes
I-Like	Managing the Internet of Things (IoT) data gathering and elaboration	Holonix
MaGA	Managing the environmental impact analysis	SUPSI
LCPA	Managing the product service life cycle cost analysis	BALANCE

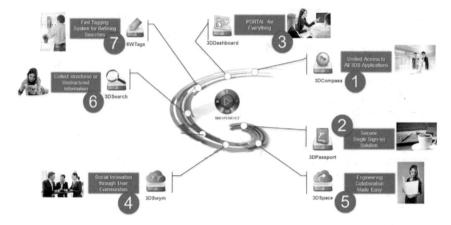

Fig. 4.2 3DEXPERIENCE's components

4.3.1 3DEXPERIENCE

3DEXPERIENCE is a Business Platform that includes Product Lifecycle Management and provides support for Business Process applications. These applications
contain pre-defined schema and processes to support many business industrial sectors (Fig. 4.2).

Industry processes and solutions leverage following 3DEXPERIENCE services:

3DCompass

It's the "key to the 3DEXPERIENCE Platform"

- The compass manages access to the applications in the ***3DEXPERIENCE***
- Each user has a personalized view and access to his/her licensed applications based
  on their selected Roles;
- Each quadrant of the compass opens a specific category of applications:

    - West: 3D Modelling (CATIA, SolidWorks).
    - South: Content and Simulation (SIMULIA, DELMIA).
    - East: Information Intelligence (3DDashboard).
    - North: Social and Collaboration apps (ENOVIA, 3DSwym).[1]

3DPassport

3DPassport provides a secure single sign-on environment for the entire 3DEXPE-
RIENCE Platform. It is based on the industry standard CAS (Centralized Authentication Service—open standard for authentication management server). In particular
3DPassport is implemented on top of CAS Server version 3.5.2.

---

[1]CATIA, SolidWorks, SIMULIA, DELMIA, ENOVIA are Dassault Systemes brands.

3DDashboard

With 3DDashboard, user can create their own dashboard for rapid, intuitive visualization of business and product data. 3DDashboard helps managers to ask the right questions and connect the dots in the Platform. Also, the dashboard can be used for *Social Media Listening* where widgets are automatically created for a specific topic.

3DSwym

3DSwym can create social communities to collaborate in an unstructured environment. Communities contain Web 2.0 collaboration tools such as Blogs, Wikis, iQuestions.

3DSpace

It's the "core" of the 3DEXPERIENCE Platform, used to manage and share content (data, documents and related information) for effective collaboration.

The main functionalities used in the Manutelligence project are:

- *Product Planning Programs*: to improve project management execution with flexible calendars, interactive Gantt edition and to monitor the project execution with new summary view and standardized reports. These functionalities are used, in particular, to monitoring the Manutelligence project itself.
- *Global Product Development*: engineering collaboration with Product Structure/EBOM integrated experience (Lifecycle and Configuration) and improve end-to-end change governance from requirements to engineering. Managing requirements traceability versus test executions and prototypes As-Built BOM. The *Ferrari*, *Meyer* and *FabLab* industrial cases have been developed based on these functionalities (Fig. 4.3).

Fig. 4.3 Dashboard of Manutelligence project inside 3DEXPERIENCE

The 3DEXPERIENCE Platform Architecture is a three-tier architecture in which the presentation, the application processing and the data management are logically separate processes. The Architecture is a web based one, using the https protocol, to ensure secure data management. Access to the application can be done via web browser (Webtop Client) or via specific client application (Native Client), as shown in the following Fig. 4.4.

4.3.2 I-like

I-Like is the Holonix web platform supporting the Internet of Things (IoT) processes in the Manutelligence platform.

I-Like Architecture

The Holonix I-Like is a web platform aimed at retrieving, organizing and visualizing all the data that are relevant to know the history and the current status of a machine or product. This is the module inside the platform whose objective is to acquire data from the field. Once the data are acquired, they can be used for different analysis and with different purposes. For example, inside the Manutelligence Platform, data coming from different "sources" like Ferrari FXX high performance race car, FabLab 3D printers, Lindbäcks wooden houses and Meyer ships are captured by Holonix I-Like. The core of the solution consists of a cloud platform, a set of gateways to read data from the field and a set of web and mobile apps to present the data to the users. Following Fig. 4.5 offers a general graphical representation of the Holonix I-Like architecture.

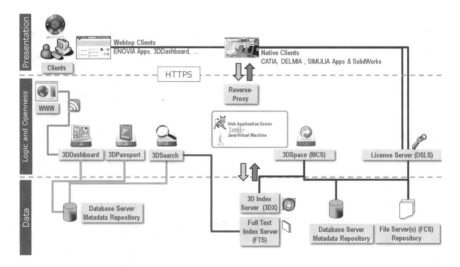

Fig. 4.4 Three-tier architecture

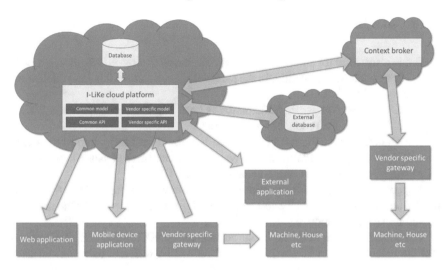

Fig. 4.5 *I-Like* architectural overview

The cloud platform is in charge of the following four main tasks:

- Storing of the relevant information collected during the machine or product life-cycle,
- Maintaining a complete representation, at any moment, of the machine or product current status,
- Keeping track of the machine or product status history,
- Detecting machines or product alarms and anomalous conditions and notify the users and maintainers about them.

To feed the cloud platform, data must be collected from the field. This is achieved by implementing a soft-ware component (gateway) that talks to the machine or product, does the basic computations that are easy to be performed with low latency access to the machine and sends the data in a secure way to the cloud platform adhering to its API. This part is often customized to the specific case, as protocols might change across various machines types and might be proprietary. It can reside on hardware already present on the machines or on embedded systems added on purpose. A Representational State Transfer (REST) API allows exchanging information with third-party applications while the Application framework is used to interconnect together many modules providing security aspects, relations and logics with the stored data.

4.3.3 MaGA

MaGA is a software for Life Cycle Assessment developed by SUPSI. It was developed as a standalone application. In the Manutelligence context it is integrated as an application of the platform.

The Life Cycle Assessment calculation is based on the availability of a Bill of Material (BOM) of the Product to be analyzed.

The Manutelligence Platform is designed in order to let MaGA import the BOM from 3DEXPERIENCE Platform. Anyway MaGA can be used standalone and the BOM can be created from scratch using the MaGA UI.

A specific interface was developed so that the user can, during the MaGA session, import the BOM automatically and add all the information needed for the environmental impact calculation.

The MaGA session data can be saved into the 3DEXPERIENCE Platform, so that the calculation can be carried out and refined by different users on different workstation as needed by the work organization.

Once the Assessment calculations are completed, the final results can be uploaded in 3DEXPERIENCE Platform, making it available to all the users, as per the access rules implemented.

4.3.4 LCPA

LCPA is a software for Life Cycle Costing developed by Balance. It was developed as a standalone application. In the Manutelligence context it is integrated as an application of the platform.

As in the case of MaGA, the Assessment calculation is based on the availability of a Bill of Material (BOM) of the Product to be analyzed, and for that reason the Manutelligence Platform was designed to let LCPA import the BOM from 3DEXPERIENCE Platform. Anyway LCPA can be used standalone and the BOM can be created from scratch using the LCPA UI.

A specific interface was developed so that the user can, during the LCPA session, import in the BOM automatically and add all the information needed for the environmental impact calculation.

Once the calculations are completed, the final result can be uploaded in 3DEXPERIENCE Platform, making it available to all the users.

Table 4.2 Data exchange

Involved systems	Technology used	Exchange format data
MaGA $\leftarrow \rightarrow$ 3DEXPERIENCE	REST web service	JSON/XML format
LCPA $\leftarrow \rightarrow$ 3DEXPERIENCE	REST web service	JSON/XML format
I-Like $\leftarrow \rightarrow$ 3DEXPERIENCE	REST web service	JSON format
I-Like $\leftarrow \rightarrow$ LCPA	REST web service	JSON format

4.4   P-S Collaborative Design and Manufacturing Platform Integration

Data exchange between four components of Manutelligence Platform (3DEXPERI-ENCE, I-Like, MaGA, LCPA) is ensured through the use of Web Services technology that provide a standard means of interoperating between different software applications, running on a variety of platforms and/or frameworks.[2] The following table shows technology end exchange formats adopted (Table 4.2).

The workflow of the data exchange between 3DEXPERIENCE and MaGA as well as 3DEXPERIENCE and LCPA can be summarized in the following steps:

- Retrieve list of existing products from 3DEXPERIENCE to MaGA or LCPA.
- Retrieve the BOM (Bill of Material) for a specific product from 3DEXPERIENCE and import it into MaGA or LCPA tool.
- Perform the Assessment with MaGA or LCPA tool.
- Send the Assessment from MaGA or LCPA tool to 3DEXPERIENCE to make it available for the platform users collaboration.

This solution will allow an iterative process to converge to the optimization of the environmental impact as well as on the life cycle costing since the early stages of the Product-Service design. In fact this workflow can be applied immediately, driving the materials selection, the design solution and taking into account the whole life cycle of the product, including the service and the disposal phases.

The workflows of the data exchange between 3DEXPERIENCE and I-Like can be summarized in the following steps.

Meyer Case

- The "Meyer management system", hosted by Holonix, is retrieving, organising and visualising all the data that are relevant in order to know the history and the current status of a specific Boat.
- When an anomalous condition occurs, the on board system launches the alert to the operator for the local action and transmits the information about anomalous condition to the 3DEXPERIENCE Platform creating automatically an Issue.
- The Issue can be the start point for the Change Management process.

---

[2]https://www.w3.org/TR/ws-arch/.

Ferrari Case

- The telemetry system, hosted by Ferrari plus Holonix (steering wheel accelerometer), is retrieving data measured during the specific trip of a specific car; the I-Like user can organize and visualizing such data.
- The organized data will be automatically transferred to the 3DEXPERIENCE to allow the structural analysis, in particular the frequency and modal analysis to improve the drive style performance by introducing the appropriate powertrain transmission changes to avoid resonance and vibration amplitude. The data transferred will be classified and identified by 3DEXPERIENCE, providing to I-Like a unique identifier of the specific trip session.

Lindbäcks apartment Life Cycle Costing Case

The workflow of the data exchange between *I-Like* and *LCPA* can be summarized into the following steps.

- I-Like gets and stores data coming from Lindbäcks apartment (e.g. temperature, alarm, humidity).
- LCPA automatically downloads from I-Like the averages measured values of the sensors for a given interval; the energy consumption will be calculated in LCPA.
- The real energy consumption, based on measurements in the apartment, and the energy costs, based on actual price models, are calculated.
- LCPA tool compares the "real" energy consumption of the apartment based on measurements with the calculated energy consumption based on mathematical models.

In this way the apartment designer is able to compare theoretical calculation with measured data and then can improve the energy consumption due to optimized isolations for the next apartment.

The apartment tenant is aware of his energy consumption and the energy consumption costs on a daily basis and also room-related (in case the measurements are done in every room) and he is enabled to adapt his energy consumption habits based on real measured data.

4.5 P-S Process Design and Manufacturing Execution Tools: 3D Modeling (CATIA and Solidworks)

The goal of the activity of the Design Engineer is to create and document the process required to develop new objects from the concept phase, up to the engineering phase. All the business processes are implemented by CATIA and Solidworks modules, in the framework of Dassault's *3DEXPERIENCE* platform.

4.5.1   Product Structure

The Product Structure is the product representation of the way it is conceived, developed and engineered. This is the basis to develop the Bill of Materials and to execute simulations to validate the product.

In the context of CATIA module the basic objects to design are:

- Physical Product, used for the assemblies (the "branches" of the product tree).
- Physical 3D Part, used for the component (the "leaves" of the product tree).
- Representations.
- 3DShape, containing the 3D geometry.
- Drawing, containing the 2D geometry.

These object are used to create the Product Structure (also know as product tree) and to create the 3D geometry, enabling the visualization experience of the products (Fig. 4.6).

4.6   3D Modeling

The design of a product can start from scratch by creating a sketch or re-using library objects or via carry-over of previous products to modify to design new ones. There is a bunch of functionalities available to capture the intent of the designer in order to facilitate and speed up the product development, as well as tool to control the data access and the product maturity follow-up in a quality context.

The results that can be obtained are the ones developed for the Manutelligence industrial use cases (Fig. 4.7).

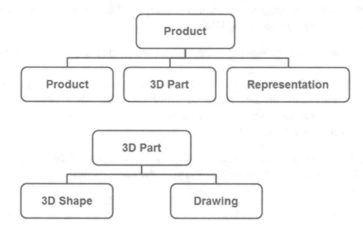

Fig. 4.6 Product's structure

Fig. 4.7 Ship example

Similar result can be obtained using the Solidworks module, part of the 3DEX-PERIENCE Platform, as done by Fundacio CIM for the FabLab case of the lamp.

4.6.1 Visualization

The visualization is supported by the 3DEXPERIENCE and by the Manutelligence Platform, leveraging on the CATIA and Solidworks features.

In particular, the visualization is useful when applied to the sessions of Digital Mock Up (DMU) usually adopted for design reviews, where designers and product managers analyze the progress of product development for approval milestones.

Typical features offered by the DMU are:

- The analysis of entire assembly, filtering as needed to check the desired parts or systems.
- The clash analysis on a configured scenario, i.e. by selecting one of the possible configurations of the product, e.g. a car with right or left driving seat.
- The visualization for consulting is available with web applications that do not require specific CAD capabilities.

As example for the FabLab case, the 3D visualization is available navigating the Bill of Materials in the web browser, offering view and annotation tools (Fig. 4.8).

Fig. 4.8   Visualization of the Moveo (FabLab case)

4.6.2   Simulation Modeling (SIMULIA)

The 3DEXPERIENCE Platform extends traditional product lifecycle management (PLM) concepts to simulation, by offering a single platform for managing CAD and simulation data and processes, thus permitting seamless design iteration based on simulation results. The Manutelligence Platform is extending these capabilities by directly being fed with real data captured during product usage thanks to the IoT solution (Holonix).

The goal of the activity of the Simulation Analyst Engineer is to simulate with virtual tools the behavior of the objects being developed by the Design Engineers to validate and document the process and to develop new objects from the concept phase up to the engineering phase. All the business processes illustrated are implemented in the Physics Modeling apps, based on the SIMULIA technology, within the framework of Dassault's 3DEXPERIENCE Platform.

Simulation data in the 3DEXPERIENCE Platform are organized into three major categories: Model, Scenario and Results.

Model

The Model contains the data required to perform a finite element simulation of the product. In particular, the Model contains a finite element model representation (FEM Rep) that is a representation of a model, always associated with either a 3D Part or a Physical product, in which the geometry is discretized into many geometrically simple elements (mesh) according to the finite element method.

A FEM Rep can be associated with a simulation object thereby allowing multiple simulations to be associated with the assembly. These FEM Reps can contain different meshes and associated properties.

The 3D Part or Product mesh can be created using several different algorithms In particular, the mesh of the 3D volume can be obtained by using tetrahedral or hexahedral elements with different levels of refinement; in the same way, the 3D surface mesh can use either triangular or quadrilateral elements. The model, in addition to the mesh elements, can include connector elements that simulate the physical connections of relative movement and degrees of stiffness values/damping, imitating the behavior between the parts of the model (Fig. 4.9).

Fig. 4.9 Finite element model using tetrahedral elements. In red color, connector elements schematizing a shock absorber

Scenario

The Scenario consists of different objects that describe how a simulation is performed; the availability of certain types of actions will depend on the type of scenario creation app.

In the Scenario, the type of analysis procedure to be performed is specified; this can be a general static analysis, a non-linear implicit or explicit dynamic analysis, or a linear dynamic analysis. Within each steps, the relevant external loads (e.g. gravity, pressure, concentrated forces) and restraints are applied to the model, together with output requests for the analysis.

In particular, in the Scenario the link between the CAD data and the Internet of Things data occurs. For example, considering the Ferrari use case, telemetric data coming from the test drive and processed by Holonix are imported onto the Manutelligence Platform in tabular format (e.g. CSV format) and used to create (amplitudes of the forces in time/shift) the applied loads and/or displacement as boundary conditions (Fig. 4.10).

Once the Simulation Scenario has been completely defined, it is possible to execute the analysis by specifying the number of cores to use to leverage the scalability of the SIMULIA solver.

Results

Results, in the context of simulation, contain the outcome of the executed analysis, including output variables, reports and animations that can be viewed in the Physics Results Explorer app.

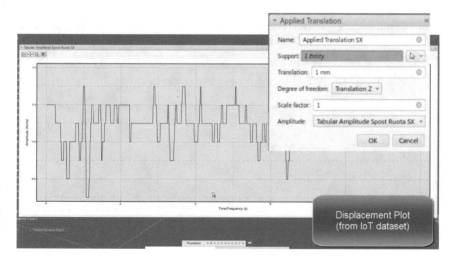

Fig. 4.10 Input from telemetry

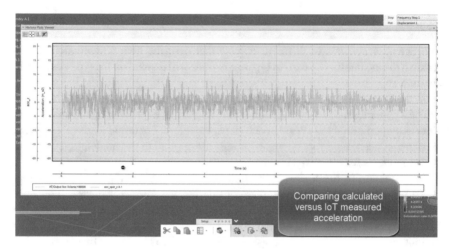

Fig. 4.11 Comparing simulation results versus measured data

The Physics Results Explorer app enables results visualization from applications relevant to both designers and experts. In addition, it enables collaboration between multiple users and integration with design optimization. Physics Results Explorer uses parallel processing to take advantage of high-performance, multi-core systems and enables to process the results of large realistic finite element simulations.

Regarding Ferrari use case, two analysis steps were performed. In the first step, the resonance frequencies of the structure were extracted; in the second step, the dynamic behavior of the model was evaluated. Then, results were post-processed in terms of resonance frequencies and corresponding deformation modes in the first step, while in the second step the dynamic forces/displacements acting on the system were plotted.

Moreover, history plots are available as X–Y curves; in this way, time-dependent quantities calculated during the numerical analysis and physical quantities available thanks to the IoT integration, are easily comparable within the Manutelligence Platform (Fig. 4.11).

Chapter 5
Tools and Procedures to Embed and Retrieve Product-Service Lifecycle Knowledge

Jacopo Cassina, Ida Critelli, Lara Binotti, Eva Coscia and Stefano Borgia

Abstract The cross-disciplinary collaborative management environment developed in Manutelligence project for Product-Service (P-S) engineering has the primary objective of managing all data, information and knowledge related to the P-S and its lifecycle in manufacturing. The present chapter presents tools and procedures to embed and retrieve this lifecycle knowledge. As extracting feedback from customers can make the P-S more suitable and attractive to the customers themselves, improving the P-S design and the development of new services, first of all an analysis of the approaches for customer feedback based on the Manutelligence industrial cases (automotive, ship, smart house, 3D-printing) has been performed. The customer feedback is characterized through different viewpoints: customer types, benefits, feedback content and methods for feedback collection, also focusing on how the Manutelligence platform can support the feedback collection. Concerning the use cases, practical examples on some users experience performed using the modules of Manutelligence platform are described. In particular, the application developed for Ferrari data retrieval and the visualization tool of Mayer are introduced as significant modules supporting interaction and feedback between the customer and designer. Taking into account the information coming from different sources (e.g. PLCs, sensorial IoT nodes, etc.) in the context of production system, a methodology to present coherently field data is proposed, with the idea to offer interoperability and integration also from the point of view of the different devices used to collect these data. For this purposed, a specific application—part of i-LiKe suite and comprising a specific dashboard for manufacturing data visualization—has been conceived. In order to explain the related capabilities and functionalities, a simulator related to 3D printing is presented as an extended demonstration of the proposed approach.

J. Cassina (✉) · I. Critelli · L. Binotti · E. Coscia · S. Borgia
Holonix s.r.l., Corso Italia, 8, 20821 Meda, MB, Italy
e-mail: Jacopo.cassina@holonix.it

© The Author(s) 2019
L. Cattaneo and S. Terzi (eds.), *Models, Methods and Tools for Product Service Design*, PoliMI SpringerBriefs, https://doi.org/10.1007/978-3-319-95849-1_5

5.1 Methodologies for Customer Feedback

5.1.1 Introduction

Extracting feedback from P-S customers can make the P-S more suitable and attractive to the customers themselves. However, it is always not easy to get the feedback. The objective in this paragraph is to analyse the approaches for customer feedback based on the Manutelligence industrial cases: automotive, ship, smart house, fablab/3D-printing. The customer feedback is analysed through different viewpoints: customer types, benefits, feedback content and methods for feedback collection. Additionally it is discussed, how the Manutelligence platform can support the feedback collection.

5.1.2 Customer Feedback Analysis Approach

The Manutelligence platform aims to enable designers and engineers access data from the traditional enterprise IT systems, but also from the IoT enabled systems. The objective is to manage all data, information and knowledge related to the P-S and its lifecycle in manufacturing. In the beginning of the project each of the four use case pilots specified the use scenarios, including descriptions of processes or process parts which could be supported by a P-S engineering platform. These scenarios were the main source for this feedback analysis. The descriptions included the objectives, challenges, lifecycle stage and the use cases included. Each scenario could have more than one use case. Each use case again was described with a common format including for example actors, precondition, post-condition, systems involved, diagram and the main steps. The number of scenarios and use cases defined were 14 scenarios and 34 use cases.

Here the Manutelligence use cases have been used as the information source to analyse real world needs, opportunities and methods for customer feedback. The analysis included the following steps:

- Identification of the use scenarios and use cases in which customer is involved as an active or passive source of feedback. As a whole in 7 of the 14 scenarios and 9 of the 34 use cases some kind of customer feedback was included.
- Analysis of the identified scenarios and use cases using a common approach and template was done.
- Consolidation of the analysis results. This was performed for the items considered most important, like the customer/user type, objectives and benefits, lifecycle stage, feedback type and channels or tools, how the feedback is used and what could be the role of the Manutelligence platform.
- Identification of similarities and differences; developing a mapping framework and conclusions.

5.1.3 Customer Types

When talking about customer feedback, it is important to identify that there are different types of customers. In Manutelligence the following definitions are given:

- Customer: "Actors (typically a person or organization) who request and pay for a value representation".
- User: "Actors who take immediate benefit from a value representation (e.g. ownership of a product)".

Sometimes the customers are the same as users but in other cases the customer is not necessarily the user: the user may be the customer of the customer, or even the customer of the customer of the customer. As an example: the customer of a shipyard is the ship owner and the final end users are the travellers. Additionally, the ship owner is also the user in the operation phase but with a different focus than the travellers. More, in addition to the owner and end user, there may be also actors who operate the P-S.

In principle the feedback from customers may come from different customer/user levels and the P-S providers would be interested to get the feedback as early as possible in the lifecycle. However, typically it is not as easy to get feedback from the user level (if not the first customer) in the lifecycle phase when the P-S is still in the design or manufacturing/implementation phase. On the other hand, in the usage phase it is often more simple to get the feedback from the user.

5.1.4 Objectives and Benefits

Even if the Manutelligence use cases significantly differ from each other in size and complexity, they all express, in different ways, one common objective for the collection of the customer feedback. It is the improvement of the P-S, either of the P-S instantiation (a specific P-S) or the future P-Ss. The improved P-S is expected to influence the customer satisfaction and thus to improve the competitiveness of the company. Additionally it is expected that the interaction including the feedback and the potential to influence the P-S design strengthens the customer relationship and enables better understanding about the future needs.

The availability of life cycle analysis (LCA) and life cycle cost analysis (LCC) on the platform allows the integration of LCA and LCC analysis into the design process. The results can be offered to the customer and the customer is able to give feedback if the P-S performance is sufficient. The end user can make the decision based on sustainability assessment and long term costs. For example, the customer can predict future energy consumption and ask for changes if the performance is not good enough.

In addition to high quality P-S, one expected benefit is to speed up the design, P-S specification and implementation processes, and to decrease the costs. Efficient feedback tools enabled by the platform allow faster fixing of the design decisions, but also avoiding errors in the design. Thus there is decreased need to waste time for the correction of errors and the subsequent manufacturing/implementation phases may be more efficient.

5.1.5 Mapping Feedback Scenarios

As described above, the main reasons for collecting customer feedback is to improve the P-S design and the development of new services. Thus the content of the feedback is mainly customer opinions and data about the P-S performance. The customer feedback could be categorized with two main dimensions:

- The P-S lifecycle phase in which the feedback is collected and analysed. Main phases identified are P-S design/implementation and P-S operation/use phases. The end-of-life phase was not visible in the scenarios.
- The type and method of feedback: type meaning the information type (unstructured information, structured information, data) and method meaning how the feedback is given (customer manual input, customer selection from predefined options, automatic (for example) sensor data). It seems clear that in most cases type and method are not independent but interlinked: the unstructured information requires some customer activity while the bigger amounts of data are coming automatically from sensors, for example via IoT. The structured feedback (for example selection between options) can be derived either from the customer or from automatic devices.

Thus the main dimensions against which the use cases can be compared and analysed are: P-S lifecycle phase and the feedback type and method; customer activity with different types of data/automatic retrieval with structured data. The customer activity may mean feedback given through the platform, email or discussions in a meeting etc. In the project has been shown that the different use cases had a different focus in their scenarios and collaboration with customers.

5.1.6 Role of Manutelligence Platform

Analysis of the Manutelligence case scenarios brings out that there is a need for retrieving customer feedback in all lifecycle phases even if no end of life scenarios were available in the current project. It is clear that an IT platform managing the design and feedback information and utilizing IoT (Internet of Things) is needed to collect and manage the large amounts of data given by automatic sensors. The data can be used for use phase services and for further design. Often there is also a need to compare the real data against designed performance (for example energy consumption models).

To receive feedback from the customer through customer actions the P-S provider needs to make the feedback action attractive for the customer. This means that it should be easy and interesting for the customer, for example to understand the current P-S version in the design phase, and also easy to give the feedback. In the use cases this was implemented through visualization, even experimenting through gamification, supported by the platform. The feedback could be directly appointed to the visual models or given by more traditional means, like in meetings. Thus also here the platform is needed both to present the P-S for which feedback is needed but also to save and analyse the collected, often heterogeneous information.

In the Fablab-case there is a user community, which is interested to interact and give feedback to the P-S provider but also to share experiences. Thus the platform needs to offer tools also for this communication and collaboration.

As an important function of the platform in relation to customer feedback is to take care that the customer feedback is handled and used for the current P-S instantiation or for future P-Ss. Thus systematic change management process, supported by the platform, is needed. The change process also takes care about informing the customer about the results of the feedback process: what changes were made.

As a conclusion, a P-S platform may have different roles in the customer feedback process:

- The platform should manage the rich P-S data and information throughout the lifecycle.
- The platform should offer the P-S information to the customer in an understandable interface.
- The platform should offer the customer a possibility to give different types of comments.
- The platform should integrate to IoT to collect data from different types of sensors.
- It should be possible to analyse the feedback data and information using the platform, for example to compare real and designed data.
- The platform should support the change management process.
- The platform should allow communication between different users or different actors.
- The platform should support organizational change management by enabling dynamic changes required by changes in roles and tasks.

5.2   Tools for Customer Involvement: Manutelligence Web Application User Interfaces

The objective of this paragraph is to provide some practical examples, related to the industrial use cases, on some users experience performed using modules of the Manutelligence platform. Not all the scenarios are presented for the sake of brevity. The next paragraphs present also different tools adopted to involve different customers inside the platform.

5.2.1 Ferrari Use Case User Experience

The objective of this paragraph is to illustrate the application developed for Ferrari data retrieval from the point of view of the user stream. These data can be used to provide feedbacks to two types of customers. More specifically, in this scenario, it is possible to provide feedback to the Ferrari designer who can use the data acquisition to improve the design of the future cars and also to the driver of the car, who can visualize the progress of different variables of the test drive performed on the track. After the login inside the application, the user can access the list of cars that have been tracked inside the platform as illustrated in Fig. 5.1. By clicking new car, it is also possible to add manually a new record belonging to the list of cars.

In the same way, also a list of gateways (Fig. 5.2), used to acquire data and associated to a specific car, is made available to the user.

The user can select a specific track associated to a specific car and then can visualize the GPS position of the car and also the variables acquired as illustrated in Fig. 5.3.

The user can create a placeholder for a specific event, e.g. it is possible to report if at some point of the test drive something relevant has happened as shown in Fig. 5.4.

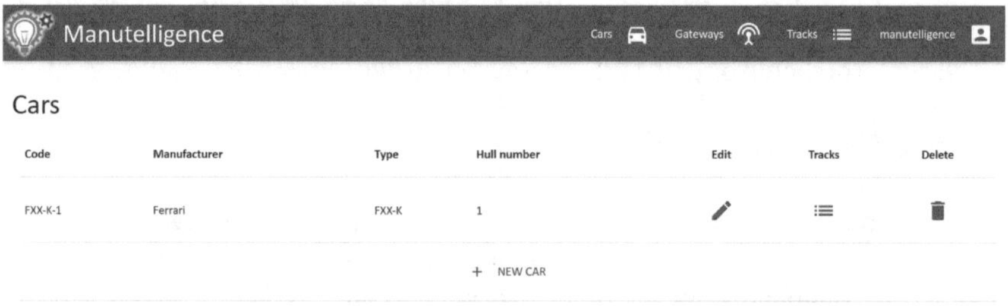

Fig. 5.1 List of cars

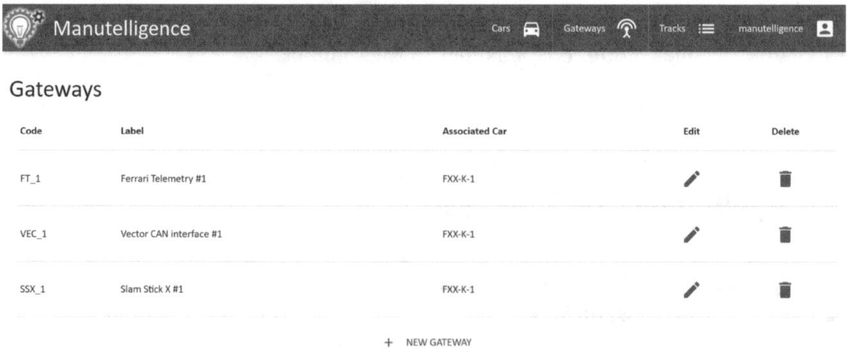

Fig. 5.2 List of gateways

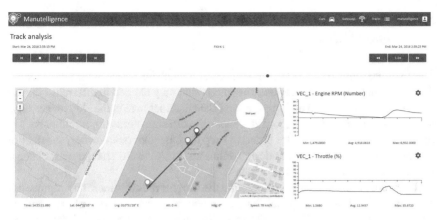

Fig. 5.3 Track data visualization

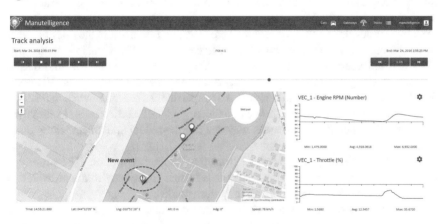

Fig. 5.4 Visualization on the map of the placeholder related to a specific event

As a conclusion, in this module of the *Manutelligence* platform, the roles in the customer feedback process exploited are:

- Offer the information to the customer in an understandable interface.
- Offer the customer a possibility to give different types of comments.
- Integrate the IoT to collect data from different types of sensors.
- Analyse the feedback data and information using the platform, for example to compare real and designed data.

5.2.2 Meyer Use Case: Customer Feedback in Sales and Design

This use case aims to support the interaction and feedback between the customer and design through ship visualization. The interaction may happen in different lifecycle phases, like sales, design and production. The customer here is the ship owner organization which may include about 200 experts in different fields (e.g. architects). The objective may be to win a sales order, to keep the customer (ship owner) satisfaction high, to get acceptance for the design solutions and to avoid change needs later in the design or in the manufacturing phase.

There are different means to modelling the ship for the customer. The engineering department can make the visual 3D ship model available to the customer and the customer can view the ship model from the engineering platform, rotate it from different angles, move in the model, and zoom in/out, but not change the model as such. A more advanced Virtual Reality demonstration using gaming technologies has been further developed and is presented here. The aim is to use it in the sales phase to make the customer convinced about the offered solution. The scenario "Immersive 3D Walking Experience" with a networked VR platform can be used with different devices, as presented in Fig. 5.5. Clients can connect to the same global VR-pool. The data security is high in the solution. All the parties can see their own personal VR experience and the communication can happen via VOIP and Avatars (Multiplayer concept). This means that the users may walk in the model and "meet" there other users.

Figure 5.6 presents a print-screen from the VR demonstration. The prototype contains a conversion of the ship model to a virtual reality environment, which is compatible with all platforms also in the future market.

The visualization can be considered as an additional knowledge-based service to the customer, which helps to sell more ships and keep the customer satisfied. The knowledge based service may include importing, cleaning and optimizing the CAD data for the game engine use, adding UI components for movements, interactive objects, creating VR-solution for various environments HTC-Vive, Hololens, advanced lighting, textures and sounds.

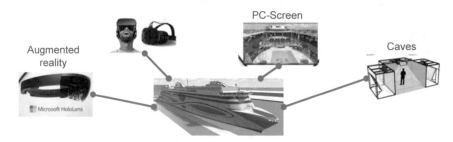

Fig. 5.5 Mock-up of networked VR platform

Fig. 5.6 A printed screen from the ship VR demo

Further development can be made towards new better people in the model (more realistic, better graphics and animations), taking into account the ship environments on where the ship will be travelling + day&night, and upgraded ship models to look good from closer distance (some modelling work will be done, for example deck furniture can be replaced with more detailed models etc.).

As a conclusion, in this module of the Manutelligence platform, the roles in the customer feedback process exploited are:

- Offer the information to the customer in an understandable interface.
- Offer the customer a possibility to give different types of comments.
- Allow communication between different users or different actors.

5.3 Manufacturing Context Driven Intelligence Layer Development and Integration

5.3.1 Introduction

The objective of this paragraph is to present a methodology to present coherently data coming from different devices. What has been designed can aggregate data coming from different devices, like PLCs or sensorial IoT nodes, which are meant to collect data mostly from industrial machines in the factory and present them in a modular and fully configurable dashboard. In order to do so, the application developed for managing different devices, which is also part of the i-LiKe suite, is described in the following. With the purpose to explain what the capabilities and functionalities are, a simulator related to FabLab machines is presented as an extended demonstration of this approach.

5.3.2   Device Abstraction Layer

The device abstraction layer (D.A.L.) is an application developed specifically to manage the devices in the environment of a given project, providing a unified point of access to device data for the other applications.

In particular, it is capable of providing an "image" of the device status built upon the data collected, applying rules to the incoming stream of data, transforming raw low level data into a more meaningful form, recording the data to long term storage and providing a way to retrieve the data recorded. It also provides access control for the devices connected, by distinguishing access for devices, gateways (e.g. a single board PC publishing data for multiple devices) and applications (which will consume the data).

Basically, D.A.L. represents a developed stand-alone application. Considering the Manutelligence platform, it is integrated inside the I-Like module, representing a middle layer between the IoT devices and the component supporting data collecting and preliminary analysis. So D.A.L. can be seen as the interchange data point between devices and the overall SW platform: it exposes a set of REST APIs that allow the dashboard to access data.

Generic implementation

To model the devices, and the rules applied onto them, three main entities are used:

- *DeviceType*: they group devices of the same type (e.g. a 3D printer model) and related rules sets;
- *DeviceTypeVersion*: the version is where the rules for a device type are really contained; it is useful as in development stage different rules sets for the same device type might be needed, so that old devices can continue working on established rules set while new ones can be experimented;
- *Device*: they are the instances of a DeviceType (e.g. the single 3D printers sold by a manufacturer), each one with its status and data.

To provide flexibility required to manage data originating from various different devices, a collection of rule "blocks" is available. The blocks can be combined with each other into the DeviceTypeVersion described above. Each rule block follows the event emitter pattern, using the base classes present in Node.JS runtime environment, so the data it process can directly contribute to the device status "image" and eventually it can be passed on to other rules for further processing.

The rules sets mapped into DeviceTypeVersions are then instantiated for single devices, each one operating on its encapsulated data set.

The DeviceType-DeviceTypeVersion-Device hierarchy, with related device and applications access control metadata, are related to a Context. Several contexts can be accommodated on the same instance of D.A.L., so that multitenancy can be achieved: in certain scenarios this is important as several smaller projects can be served by a rather complex setup of multiple application and databases management systems.

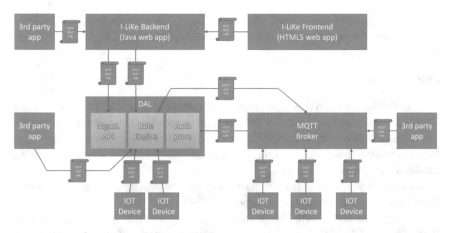

Fig. 5.7 A printed screen from the ship VR demo

The interaction and the interoperability with the system are provided through a REST API (Fig. 5.7). Basically, two sets of REST APIs are used: the first set provides REST HTTP API with JSON data encoding for the management of the devices inside the rule engine (device registry, creation of *DeviceType* and *Device*, activation/deactivation of data acquisition and elaboration, generation of new credentials, etc.); the second one makes available APIs to interact and visualize time series monitored data, alarm list, etc.

Data sources

The first main task of the D.A.L. is to provide contact points where the device can send its data. Two protocols are supported to bring the data from devices onto the system:

- MQTT: (Message Queue Telemetry Transport) is a lightweight protocol based on the publish/subscribe pattern over TCP/IP protocol. MQTT requires a broker to be present, distributing messages to interested subscribers based on the topic of the message. D.A.L. provides a rule block that supports connecting to a broker and subscribe onto a topic.

 Also D.A.L. has specific support for Mosquitto MQTT broker (the most widespread open source MQTT broker), providing an interface to enable access control on it.

- HTTP: (Hypertext Transfer Protocol) is the most widespread protocol of the World Wide Web. D.A.L. provides a rule block with the capability to dynamically register a HTTP endpoint where data can be sent with common POST or PUT HTTP requests.

 Data collected by the data sources rule blocks is then emitted to the subsequent rules for manipulation.

Closely related to data sources rule blocks, D.A.L. provides buffering rule blocks, enabling time sorting of the data and cadenced emission to the subsequent elaboration pipeline.

Data manipulation

Data incoming from devices sometimes is in a ready to be stored form, but often it needs an elaboration step: simpler devices tend to provide data that greatly benefits from a stream processing approach to aggregate them into a more usable form. Other frequent use cases are the generation of a notification (event or alarm) upon some condition recognized on the data.

D.A.L. provides a set of rules to carry on these tasks in a simple manner, reducing the burden of specific implementations.

To cope with unexpected elaboration steps, the possibility to evaluate a generic Javascript function block onto the data is provided through a dedicated rule.

Data storage and streaming

For data storage several rule blocks are provided, supporting different databases and storage forms:

- Time-JSON: a generic JSON payload associated with a timestamp.
- Time-Decimal: a decimal value associated with a timestamp.
- Interval: an event data structure, identified by a UUID v4 and presenting a start date, an end date and a string value.

For long term data storage, mainly Apache Cassandra and MySQL are supported. Support to other DBMS is planned to be added by providing dedicated rule blocks.

Similarly to data storage, data streaming is also supported. Data can be streamed or stored when it is received onto the elaboration pipeline. Scheduled data extractions can also be defined.

Streaming supports sending data in the following manners:

- MQTT: by connecting to a broker and publishing on a topic.
- HTTP: by sending a POST or PUT request to a URL.
- Redis PUB: by issuing a PUBLISH command onto a Redis instance, using it as a system bus.

Data streaming is a powerful concept, potentially enabling the binding of several instances of D.A.L. in various environments: for example, a D.A.L. instance could process low level data near the devices (edge computing) and then send them for further processing and long term storage to another instance of D.A.L. on an internet accessible server (cloud computing).

Data retrieval

The last important task carried on by D.A.L. is providing a way to access the data it collected so that other applications can consume them. Similarly to HTTP data sources, dedicated rule blocks are present to dynamically register HTTP endpoints where data can be accessed through HTTP REST API call requests.

Rule blocks supports accessing the current device "image" or accessing the data stored by the data storage rule blocks.

Streaming techniques have been implemented to access the data stored in an efficient way, avoiding abnormal resource consumption with high amount of data.

5.3.3 Manufacturing Data Visualization Through Application Dashboard

To accompany the great flexibility achieved in device data management with D.A.L., an application has been developed to consume the data collected into specific domains. It is mainly a HTML5 application supported by Java based backends exposing REST APIs.

The application maps machine types (e.g. a 3D printer model) and machines (e.g. a single 3D printer); machines access device data exposed by D.A.L. through HTTP interface. Alarms are notified immediately by subscribing on a Redis channel, where D.A.L. streams alarm data.

For example, Fig. 5.8 illustrates the four typical FabLab machines that have been mapped into the application by simulating their ehavior via software as explained in the following.

Machine monitoring

A monitoring section has been implemented, focused on building a dashboard to show the relevant device data. The dashboard is built around a machine type, by composing into a web based editor, a series of panes and widgets. Widgets access

Fig. 5.8 Monitoring—machine list page

Fig. 5.9 Monitoring—machine dashboard

data from data sources, deriving by how the machine type was defined. In fact, some requirements are posed on which endpoints are exposed by D.A.L. to have a minimum set of available functionality. For example, it is possible to fully create and configure the machine dashboard.

Once the dashboard has been created and configured for the given machine types, all the machines, belonging to the same machine types and registered into the application, will be able to use them and to show their own data on it. For instance, it is possible to see the status (working, stopped, etc.) of the machine, number of cycles, power on time, laser on time and other relevant available parameters, as illustrated in Fig. 5.9. Moreover, dedicated user interfaces are provided to browse the machine alarms history and to plot and compare decimal time series charts.

Maintenance

When a machine type is registered, a maintenance plan can also be defined. The plan is composed of maintenance types, each characterized by several parameters (elapsed time, or counters on the machine status, which is mapped onto D.A.L. device "image") with scheduling parameters.

Typical scenarios on which the schedule is based are:

- Maintenances based on the time elapsed and effective working time of the machine.
- Maintenances based on the number of working cycles of the machine or on the amount of material processed by the machine.
- Maintenances based on the number of triggered events of a certain class.

The user interface provide an immediate indication of the maintenance status by showing the "health" of the machine as the lowest indicator of the remaining life according to the maintenance table. For example, as captured in Fig. 5.10, a maintenance related to the laser head replacement has been created, based on the time elapsed.

As the "health" associated to the various maintenances decreases, it is possible to schedule servicing interventions (Fig. 5.11).

Fig. 5.10 Maintenance—maintenance status

Fig. 5.11 Maintenance—servicing scheduling

It is then possible for the maintainer to clear them out, writing down notes associated to the actions performed on the machine. As the servicing is completed, the health indicators for the associated maintenance are reset.

5.3.4 Industrial Scenario: 3D Printing Monitoring

The aim of this paragraph is to present tools used for the implemented demo scenario, which was meant to describe the possibilities of IoT within the manufacturing scenario. To enable practical tests, we choose a "demo" based on 3D printers and other production machines developed by CIM-UPC, considering also the fact that CIM-UPC 3D printers are manufactured by other 3D-printers (Fig. 5.12).

Another important concept emerged working with 3D Printers, which proved to be true also for many industrial machines, as investigated during the industrial visits, is the unavailability of data from the PLC because there is no PLC (like in the scenario

Fig. 5.12 3D-printers
making 3D-printers
components in fundacio CIM

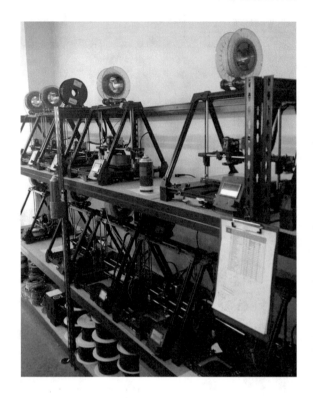

of 3D printers), or because it doesn't allow the reading of data (old or proprietary PLC). From this need, the development of a more advanced sensor node was decided.

The requirement were to have a standalone system to equip machines without a PLC or revamping older/not connected machines and that this system was going to work through an independent infrastructure, therefore not requiring a Wi-Fi in customers facilities, but only a single network cable.

From these specifications, the IoT sensor node was developed, being able to read analog and digital inputs and transmit them through MiWi, so through a different infrastructure than the WIFI, to a single data gateway, which connects all the sensor nodes to the cloud.

In the Fig. 5.13, it is possible to see a typical scenario of usage of the IoT sensor node, with 3 machines without possibility of connection to the PLC (as most of the currently on the market machines, due more than to technical issues, to cost of licenses to acquire PLC schemas).

From the machines data are read through analog connections; the IoT sensor node transforms them and sends them in MiWi to the IoT gateway, which, through a cabled connection transfers them to the cloud.

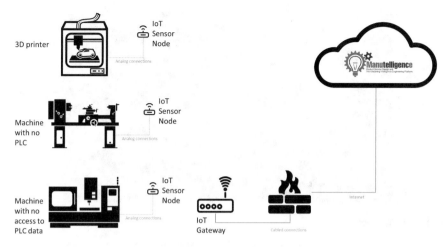

Fig. 5.13 Schema of the IoT sensor node usage

Fig. 5.14 IoT sensor node

IoT sensor node

For interfacing with machines normally unprovided of a PLC/controller or where interfacing with the machine on board controller is not sustainable, an IoT sensor node, named KISS, has been implemented to interact with the machine electrical signals, such as the ones of status lights, or by equipping simple sensors.

The sensor node is illustrated in Fig. 5.14 and it is based on a 16 bit MCU. Five isolated digital input are available (up to eight). Different signals can be identified: ON/OFF, pulses and counter. Three analog input with a 12 bit resolution and ± 10 V range are available. They can be converted into digital inputs.

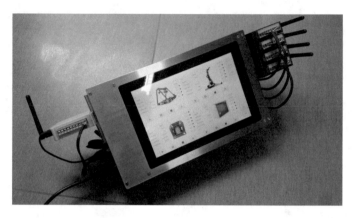

Fig. 5.15 Machines simulator

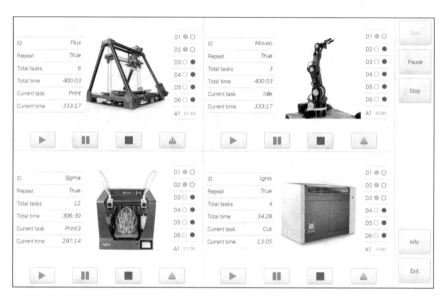

Fig. 5.16 Simulator interface

The communication is enabled by a MiWi (802.15.4) link; sensor-nodes communicates to a concentrator (hub) node, which in turn route the data collected to a gateway, usually a single board PC, where the data is finally published to MQTT through the provided MQTT client (available for both Windows and Linux operating systems).

3D printers simulation

For simulating 3D-printers behaviour, but also other types of machines, a device capable of generating electrical signals as the ones that would be available on a 3D printer has been engineered (Fig. 5.15). This device is based on a Raspberry Pi 3 single board PC, where a custom made I/O board has been interface to the SPI (Serial Peripheral Interface) bus.

An application has been then developed to control the I/O board, capable of pulling up and down digital outputs and of generating simple patterns over the analogic output. The patterns are combined into a program, which is defined into a JSON file.

The user interface (Fig. 5.16) presents up to four machines, with the ability to start, pause and stop their program or putting them in alarm state. All these events and data are available in the dashboard application and they can be easily consultable.

Chapter 6
Life Cycle Assessment and Life Cycle Costing for PSS

Donatella Corti, Alessandro Fontana, Michele De Santis, Christian Norden and Reinhard Ahlers

Abstract The increasing awareness towards sustainability issues from both practitioners and customers makes it necessary to adopt a lifecycle perspective since the design phase of PSSs. In this chapter, a tool aimed at carrying out the Life Cycle Assessment (called MaGA) and one for the Life Cycle Costing (called BAL.LCPA) are introduced starting from the analysis of requirements carried out to make sure their use is suitable in a PSS design context. In order to seamlessly include environmental and economic considerations into the design process, the two stand-alone tools have been integrated with the Manutelligence design platform. Their application in a Fablab-like environment is described to show how they interact with design tools and to provide examples of the results they get.

6.1 Introduction

The holistic approach promoted by the Manutelligence platform for the design of product service systems (PSSs) integrating a suite of collaborative tools adopts a lifecycle perspective towards a more sustainability-aware design process. In this context,

D. Corti (✉) · A. Fontana
University of Applied, Sciences and Arts of Southern Switzerland, Via Cantonale, 2C, 6928 Manno, Switzerland
e-mail: donatella.corti@supsi.ch

A. Fontana
e-mail: alessandro.fontana@supsi.ch

M. De Santis
Rina Consulting S.p.A. Materials, Technology & Innovation, Via B. Ravenna 4, 73100 Lecce, Italy
e-mail: michele.desantis@rina.org

C. Norden · R. Ahlers
BALance Technology Consulting GmbH, Contrescarpe 33, 28203 Bremen, Germany
e-mail: christian.norden@bal.eu

R. Ahlers
e-mail: reinhard.ahlers@bal.eu

L. Cattaneo and S. Terzi (eds.), *Models, Methods and Tools for Product Service Design*, PoliMI SpringerBriefs, https://doi.org/10.1007/978-3-319-95849-1_6

83

the Life Cycle Assessment (LCA) and Life Cycle Costing (LCC) tools developed for the Manutelligence platform enable the calculation of the expected environmental impacts and economic measures characterizing a certain design concept before the product/service is actually produced. In the current industrial practice, these tools are mainly used ex-post to assess the actual impacts generated by a product or a process; whereas the concepts of integrated LCA and LCC tools into the platform allow to compare, in real-time, alternative PSSs concepts also on the base of their sustainability impact. Therefore, the LCA and LCC tools have to seamlessly communicate with other databases that provide the necessary input data for the evaluation and that, in turn, could use the obtained assessment in their procedures. In this chapter, a brief state-of-the-art of LCA and LCC tools in the field of PSS is presented before introducing the tools developed in the Manutelligence context along with examples of their validation with project pilots.

6.2   Life Cycle Assessment (LCA) for Product Service Systems (PSS)

LCA quantifies all relevant emissions and resources consumed and the related impacts on environment, human health and resources that are associated with any good or service. The main reference to carry out a Life Cycle Assessment (LCA) is the ISO 14040:2006 family of standards [7, 8]. They provide a set of international guidelines internationally recognized and used as reference at global level. The identified and renowned structure of an LCA study is organized into four phases:

Definition of the goal and scope of LCA. It defines why the LCA is performed, the possible applications and other preliminary elements needed as a basis for the study such as the functional unit, the system boundaries or the allocation procedure.

Life Cycle Inventory (LCI) analysis. The exchanged natural elements between the eco-sphere and the tecno-sphere (the system analyzed), thus the resources entering (e.g. raw material, energy and ancillary material) and those leaving the system of interest (e.g. emissions, waste, products and co-products) have to be identified and quantified. This step involves data collection from several different actors and related processes along the supply network.

Life Cycle Impact Assessment (LCIA). It is meant to calculate the impacts and the effects on the environment generated by the identified LCI data.

Interpretation. This is the final phase, where the report with the quantified impacts is prepared and the critical review of the LCA results is performed.

The ISO 14040:2006 [7] standard has been developed with a physical product focus and, even though services are conceptually considered, the PSS concept is not considered explicitly. For this reason, the ISO-based LCA approach might not be applied to the PSS context directly. The main difficulty when carrying out an

LCA for a PSS is how to integrate the service component into the LCI [4]. In litera-ture, the benefits of methodologies to perform the assessment of PSSs are frequently described (see for example [2, 13, 17], whilst more rarely contributions describing how to effectively carry out these evaluations can be found [6, 10]. Often, the sus-tainable design of solutions focuses mainly on the physical product and potential optimizations are directed towards its physical subsystems, whilst only later services are paid attention. This procedure is not due to a lack of methodologies but rather due to lack of system thinking [2]. Many contributions dealing with sustainability assess-ment of PSSs depict the maintenance like the only service type [11] and, typically, even if a few methodologies are proposed to assess the environmental impacts [1, 15, 16], what they propose mainly refers to a specific type of PSS and cannot be easily generalized. The lack of general procedures that could be applied to any type of PSS could be due to the wide range of services that can be combined with a product. As a consequence, the complexity and heterogeneity of the systems that represent the PSS challenge the development of a method to systematize the information collection. Corti et al. [3] propose an approach to support the LCI phase aimed at formalizing the integration of information related to the service part of the offer.

6.3 Life Cycle Costing (LCC) for Product Service Systems (PSS)

Life Cycle Costing (LCC) is an accounting method that considers every cost flow throughout the lifecycle of a product (as defined in ISO 15686:2008 [9]).

Life cycle costs [14] can be divided into three categories: development costs, util-ity/service costs and recycling/reprocessing costs. Similarly to the cost, also revenues are allocated to the individual phases of a Product System: design phase, usage phase and recycling phase.

The results of the life cycle cost analysis are also used to optimise the design within an improvement cycle. Niemann et al. [14] have identified possible uses, such as calculation of total costs for products; identification of cost and revenue drivers; impact on outsourcing decisions; analysis of "what if" scenario or analysis of customer lifetime value.

A comprehensive review of the literature on PSS has revealed that currently no quantitative methodologies exist to assess the economic potential of a PSS [18]. Datta and Roy [5] explain that methodologies to calculate the LCC for a PSS diverge depending on the PSS model, since the estimation techniques depend on the kind of service-orientation of the system. Considering the estimation of costs, the main differences between different kinds of PSSs are the hidden costs, that cannot be quantified with traditional cost estimating methods and are due to the intangible nature of services.

Van Ostaeyen et al. [18] suggest a methodology to calculate the Life Cycle Costs for a result-oriented PSS. The methodology follows the main steps of the environmental assessment, beginning from goal, scope and definition of a unit of functional delivery. Mannweiler et al. [12] provide a step-wise procedure to calculate the LCC (and in particular Life Cycle Cost Indicator, LCCI) with the final aim of choosing the most appropriate PSS variant. They state that the only way to calculate the exact LCC is to collect the detailed information of the lifecycle characteristics that are used to describe the PSS-application, yet no methodology to correctly get this information is suggested and, in particular, there is no mention to the service part of a PSS. In order to compete in a transformed environment, companies need to properly assess the cost of their service offerings to stay competitive [5]. A classical example of application of the LCC on a service regards the maintenance. The maintenance service includes direct labour, materials, fuel, power, equipment and purchased services.

6.4 Definition of Requirements for LCA and LCC Tools

In engineering activities, such as the development of software tools, the requirements analysis is often the first step in the system design process and development, in which user's requirements are gathered and analysed to generate the corresponding tool specifications. Requirements have to be documented, actionable, measurable, testable, traceable, related to identified business needs or opportunities, and defined at a level of detail sufficient for system design. The requirements analysis carried out in Manutelligence for LCA and LCC tools had a twofold aim: first, to list the requirements these tools should satisfy in order to comply with the integration needs of the Manutelligence platform and the use for the design process; second, to decide whether to adopt existing tools (available in the market) or to develop new solutions that could better fit with the project needs.

Conceptually, requirements analysis included three types of activities:

- **elicitation**: requirements are gathered through interviews and brainstorming sessions involving different stakeholders;
- **analysis**: identified requirements are analysed to make sure they are clear, complete, consistent and unambiguous;
- **prioritization**: requirements are weighted and scored to distinguish between crucial requirements to fulfil the basic functionality and additional features.

Since technical expertise was required as well as knowledge of the Manutelligence platform features, the project partners involved in the development and integration of LCA and LCC tools have elicited the first list of requirements.

The list has been then refined through some iteration of discussions and revisions involving not only the software developers and experts, but also users represented by the pilots participating to the project. In particular, the involvement of users has been fundamental for the prioritization of requirements.

The requirements have been split in two categories: Global Requirements (coded as GR1 to GR27 in Table 6.1) and Phase requirements (coded as PR.B.1 to PR.D.6

in Table 6.1). Global requirements concern the software functionalities and their integration with the Manutelligence platform. Phase requirements are focused on the specificity of the analysis and look at features related to calculation and presentation of results. For sake of clearness, the Phase requirements have been further clustered according to the analysis phase they refer to: (i) Life Cycle Inventory (LCI) (coded as PR.B.1 to PR.B.5); (ii) Life Cycle Impact assessment (coded as PR.C.1 to PR.C.4) and (iii) interpretation of results (coded as PR.D.1 to PR.D.6). The final list of requirements used to develop the tools including the indication of their priority is shown in Table 6.1.

In order to make easier the integration of the LCA and LCC tools into the Manutelligence platform and to adapt their use to the design process, the use of GaBI (the most widespread commercial LCA tool to assess environmental impact) has been excluded since it has been considered not flexible enough. It has been decided to extend the functionalities of two proprietary tools internally developed by the project partners, namely MaGA (Manutelligence Green Application) tool for LCA and BLA.LCPA for LCC. Working on their existent versions, they have been extended in order to cover as best as possible the elicited list of requirements.

6.5 The LCA Tool: MaGA

This section describes the MaGA (ManuTelligence Green Application) tool, the software that in the Manutelligence Platform is meant to perform LCA for PSSs. According to the Manutelligence needs, MaGA allows the performing of real time analysis aimed at improving the PSS design thanks to the possibility of evaluating alternative product or process configurations from the environmental point of view. To better manage the tool's complexity, a modular approach has been adopted. The software is therefore made of many modules, each one providing specific functionalities and user-interfaces. Figure 6.1 shows the MaGA architecture and its main components.

Main components providing the functionalities needed to allow a user to carry out the LCA analysis are: *Global Editor*, *Project Editor* and *Operation Editor*.

Global Editor. It allows to edit the data needed for the sustainability assessment that concern the company's supply chain actors (such as the list and the impacts of the materials/operation used, the transportation distances or the supply chain partners location) that are involved in the whole PSS life-cycle and the set of materials and operations currently used by the company. Through this module, it is possible to introduce new data (e.g. adding a new supplier to the supply chain or to add one indicator type) or update the existing one. This data are inserted into the platform once and can be then exploited every time a new assessment is carried out. This avoids repeating the data entry process of company-specific information that are common for all the PSS projects.

Table 6.1 Final list of requirements for the LCA and LCC tools (P = Primary; S = Secondary)

Requirements
(GR.1) Easiness of integration in the Manutelligence platform (P)
(GR.2) Ability to generate real time assessment which constitutes input for design of product/processes (P)
(GR.3) Tool(s) should have programming interfaces towards other systems (P)
(GR.4) Tool capacity to adapt to platform requirements (P)
(GR.5) Integration of both LCA and LCC methodologies (P)
(GR.6) Software maintenance/upgrades available (S)
(GR.7) Intuitive user interface (S)
(GR.8) Client server architecture, within client (S)
(GR.9) Capability to deal with Product Service Systems as the object of the analysis (P)
(GR.10) Adaptability to specific ISO standards (i.e.: ISO 14025 for Environmental Product Declaration) (S)
(GR.11) Compliance with PEF/OEF Recommendation (2013/179/UE) (P)
(GR.12) Quality review instruments of modelled processes also between different locations (S)
(GR.13) Social aspects evaluation (S)
(GR.14) Persistence data backup (S)
(GR.15) Possibility to perform concurrently different activities (S)
(GR.16) Ability to support benchmark analysis and comparisons between alternatives (S)
(GR.17) Easily extensible to different industrial sectors (S)
(GR.18) Ease of deployment (multi-platform, simple installation, etc.) (S)
(GR.19) Low user skills & knowledge requirements (S)
(GR.20) Use efficiency (average time required to model a scenario) (S)
(GR.21) Cooperative multi-user capability (S)
(GR.22) Allow to perform the assessment both during the design phase and on already existing products-service systems (S)
(GR.23) Consistency check (with alarm in case of discrepancies) (S)
(GR.24) Performance of running complex LCPA (S)
(GR.25) Consider dynamic timelines like operation cost increase throughput the lifecycle (S)
(GR.26) Direct comparison of different objectives (numerical and visually) (S)
(GR.27) Comparison of different future scenarios like different fuel price development (S)
(PR.B.1) Integration of specific database (i.e.: Worldsteel, Ecoinvent, ELCD) (P)
(PR.B.2) Management of allocation procedures (S)
(PR.B.3) Possibility to modify inventory references/insert specific documents for datasets included in available database (S)
(PR.B.4) Product- Service Life Cycle modelling aligned with the designer's needs (S)
(PR.B.5) Processes parametrization (S)
(PR.C.1) Pluggability of new impact assessment methods/models (P)

(continued)

Table 6.1 (continued)

Requirements
(PR.C.2) Evaluation of specific quantities (P)
(PR.C.3) Evaluation of LCC-related indicators (P)
(PR.C.4) Possibility of structuring and analysing the results considering the different contribution of the calculated impacts (S)
(PR.D.1) Capability of creating in automatic way reports with LCIA results (S)
(PR.D.2) Sensitivity Analysis (S)
(PR.D.3) Monte Carlo Analysis (S)
(PR.D.4) Availability of normalization factors (S)
(PR.D.5) Possibility of using a customized set of indicators (S)
(PR.D.6) Different views for different users and objectives (different modes) (P)

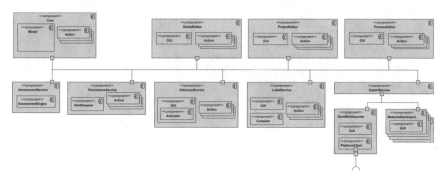

Fig. 6.1 MaGA software architecture

Project Editor. It is used ever time the assessment of a new product/service is started and project specific information (i.e. the Bill of Material or the specific set of operations) need to be edited (see Fig. 6.2). It supports the modelling of the whole PSS life cycle: design, purchase and production of sub-assemblies/components; final assembly and delivery of the PSS, the PSS middle of life (use, maintenance…) and its end of life. It is worth of notice that MaGA has the possibility to directly import information from external database, such as ECOINVENT, that provides impacts of elementary operations, thus supporting the overall calculation.

Moreover, the Project Editor provides the user with a real time calculation of the environmental indicators. Results are presented in a table-like form or with graphs and can be analysed with different level of aggregation (impact of the whole product, of a single phase or of a single components). Further, there is the possibility to compare impacts of different versions of the same product when some elements, materials or suppliers, change. Examples of impacts obtained with MaGA are shown in what follows.

Operation Editor. It directly supports the user during the design phase of PSS allowing to model the PSS lifecycle and create customized operations that are not available in the General Editor database and are not even available in existing database. Figure 6.3 shows how the Operation Editor has been used, for example, to model the 3D printing process.

The integration of MaGA with the Manutelligence platform enables the following activities (Fig. 6.4):

- import of the bill of material (BOM) from the CAD;

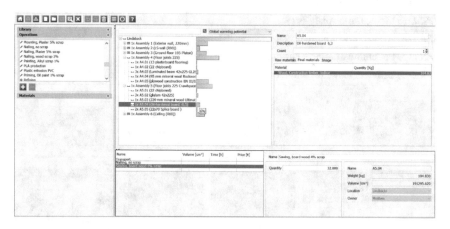

Fig. 6.2 Snapshot of the Project Editor of MaGA

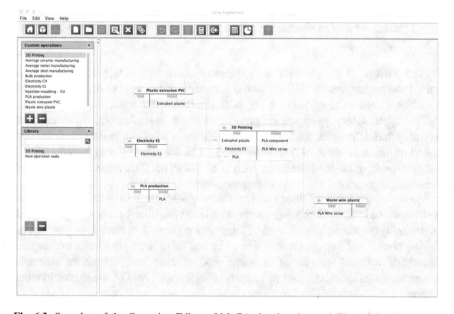

Fig. 6.3 Snapshot of the Operation Editor of MaGA showing the modelling of the 3D printing process

- export of the assessment results and possibility to carry out query of them directly in the 3D experience;
- import and export of the MaGA project files from/to the 3DExperience platform;
- import and export from/to zip files.

6.6 Testing the MaGA Tool in a FabLAB Environment

The MaGA tool has been tested by carrying out the LCA analysis for 3D printed products in the FabLAB facility based in Barcelona participating to the project. In particular, the environmental impact of a table-lamp (shown in Fig. 6.5) has been evaluated.

First, the BOM created in the CAD has been imported in MaGA and then the missing data have been inserted manually. The impacts have been calculated also

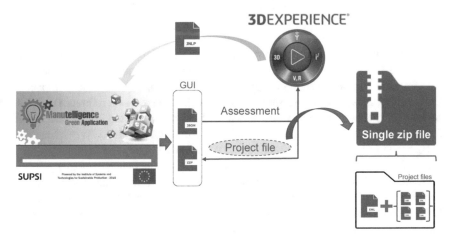

Fig. 6.4 Integration features between MaGA and the 3D Experience Testing the MaGA tool in a FabLAB environment

Fig. 6.5 The 3D printed lamp used to test the MaGA tool

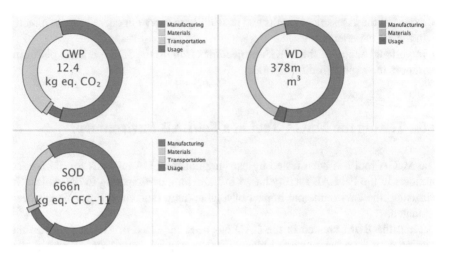

Fig. 6.6 Graphical representation of environmental impacts got with the MaGA tool when two scenarios are compares against each other

considering the use of alternative materials (PLA vs. ABS as printing material) and alternative energy mix changing the country where production is based (Spain vs. Switzerland). Figure 6.6 shows the graphical representation of the relative impact of the two alternative scenarios: Spanish mix and PLA (taken as a reference) against Swiss energy mix and ABS. Three indicators have been selected in this case out of the 11 available in MaGA, namely the Global Worming Potential (GWP), measured in eq. kg od CO_2, the Water Depletion (WD), measured in m^3, and the Stratospheric Ozone Depletion (SOD), measured in eq. kg of CFC-11, an ozone depleting gas. For each indicator, the total impact is split into the contribution of the single lifecycle phase (Manufacturing, Materials, Transportation and Usage). For example, for the GWP indicator, the graph shows that the impact on climate change generated by the production of the polymer (thus affecting the Materials phase) is higher when ABS is used, but the Usage phase impact is lowered if energy consumption is evaluated with the Swiss mix instead of the Spanish mix since nuclear power and hydroelectric power that characterize the majority of the Swiss mix have a low carbon footprint.

While the graphical representation provides an immediate idea of the impact variation of different scenarios, the tabular representation of the indicators calculated provides the precise quantification of the impacts and the corresponding percentage variation moving from the reference scenario to the alternative one. Figure 6.7 shows the results for the two-abovementioned scenarios and in the % column it reports how much an impact varies (for example, the Abiotic Depletion Potential is the 98.252% of the reference value when the ABS is used as material passing from 0.08 to 0.079 kg eq. Sb).

Since not all designers are LCA experts, in particular in a context like a FabLAB, the use of a summary label translating the impact of the assessed product into the

Indicator	Value	Reference	%	Unit
▶ ▨ Abiotic depletion potential	0.079	0.08	98.252	kg eq. Sb
▶ ▨ Acidification potential	0.085	0.105	80.31	kg eq. SO₂
▶ ▨ Endpoint total	1.935	2.303	84.035	points
▶ ▨ Eutrophication potential	0.039	0.052	73.65	kg eq. PO₄
▼ ▨ Global warming potential	12.481	13.936	89.559	kg eq. CO₂
○ Manufacturing	7.399	7.399	100	
○ Materials	4.436	3.685	120.379	
○ Transportation	0.036	0.036	100	
○ Usage	0.611	2.817	21.678	

Fig. 6.7   Tabular representation of results in MaGA when two scenarios are compared against each other

equivalent impact of simple examples, like km travelled by cars in order to represent the burden generated by the emissions of CO_2 (see Fig. 6.8), has been developed.

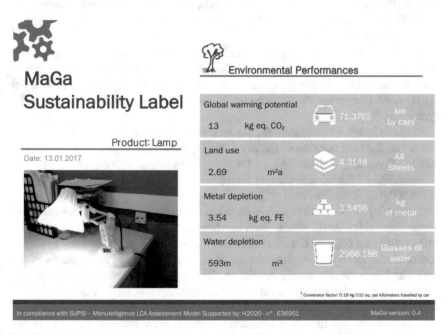

Fig. 6.8   Sustainability summary label generated by MaGA

6.7   The LCC Tool: BAL.LCPA

In order to satisfy the LCC requirements of the Manutelligence platform, the existent BAL.LCPA (Life Cycle Performance Assessment) tool has been adapted. The main economic KPI of the LCPA is the net-present value (NPV), which considers every cash flow throughout the life cycle and discounts the cash flows according to the specific point in time of their occurrence. Since the NPV accounts cash outflows and cost inflows, the flexible approach satisfies the need of determining the life cycle costs and enhances the functionality to a full investment assessment, if the user requires it. The comparable approach allows the direct investigation of different investment opportunities against each other.

A meaningful result of the LCC analysis relies on the quality of the input data. As a part of the BAL.LCPA adoption process, the tool is able to connect to the Manutelligence database in order to retrieve the required data to perform a life cycle analysis. Therefore, a dedicated interface has been developed to ease the data import, as depicted in Fig. 6.9. The data import is getting translated into a basic LCPA model, in which the different cost items and their associated cost type are assigned (Fig. 6.10). The user has the chance to modify the basic LCPA model according to his/her assessment needs, like adding an additional life cycle phase or adding additional cash flows.

In addition, the user can determine an individual cash flow development for each cash flow type. Thereby, the tool allows the generation of a cash flow timeline with a fixed annual growth rate or an individual cash flow development to reflect, for instance, an over proportional increase of maintenance costs throughout the life cycle.

In the "Global Values" section, the user can determine the NPV interest rate. Moreover, the user determines the price developments for certain global costs categories, like energy prices. BAL.LCPA also allows to consider external costs. The corresponding external cost rate per ton of harmful emissions, like CO_2, can be set in the "Global Values" as well.

The life cycle cost results are presented in tabular form as well as bar chats and curves throughout the lifecycle. Each result representation focuses on the comparison of different objects, like different design alternatives, to support the life cycle cost analysis. Each visualisation can be customised according to the needs of the user. The life cycle cost results depend on the input data as well as on the assumed circumstances of the considered lifecycle. Therefore, BAL.LCPA offers a sensitivity analysis to test the robustness of the LCC results, when certain input parameters vary. In this way, the impact of a significant energy price increase on the LCC can be tested and analysed (see sensitivity example in Fig. 6.11).

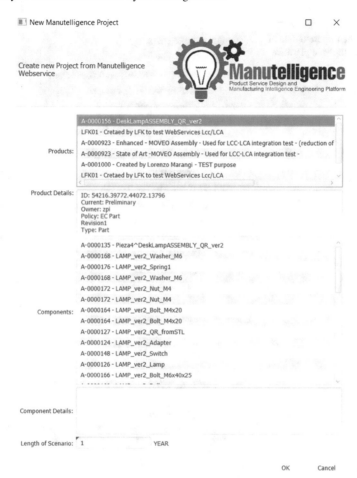

Fig. 6.9   Data import from the Manutelligence platform into BAL.LCPA

6.8   Testing the BAL.LCPA Tool in a FabLAB Environment

The described workflow to perform a LCC calculation within the Manutelligence Platform has been tested in a practical example for the FABLAB use-case: the same 3D-printed lamp used for the LCA analysis (Fig. 6.5).

To trigger the LCC analysis, the tool can be started directly from the dashboard of the Manutelligence platform. In the next step, the bill of material of the 3D-lamp stored in the Manutelligence platform, containing also life cycle cost information, is imported into BAL.LCPA. The imported data basically represents the production costs of the 3D-lamp in the beginning of life phase of its lifecycle. After the import, a simplified use-phase has been assumed to demonstrate the opportunities of a full life cycle cost analysis. Thereby, the assumed use-phase comprises a

Fig. 6.10 Life cycle model based on the imported data of the Manutelligence platform

▼ 🖳 Lamp comparison	
▼ ◯ 3d printed lamp (LED)	
▼ 🗂 Production phase	
☐ A-0000135 - Pieza4^DeskLampASSEMBLY_QR_...	
▼ ☐ A-0000168 - LAMP_ver2_Washer_M6	
🗋 Procurement costs	0.13 EUR
▼ ☐ A-0000176 - LAMP_ver2_Spring1	
🗋 Procurement costs	2 EUR
▼ ☐ A-0000168 - LAMP_ver2_Washer_M6	
🗋 Procurement costs	0.13 EUR
▼ ☐ A-0000172 - LAMP_ver2_Nut_M4	
🗋 Procurement costs	0.28 EUR
▼ ☐ Copy of A-0000172 - LAMP_ver2_Nut_M4	
🗋 Procurement costs	0.28 EUR
▼ ☐ A-0000164 - LAMP_ver2_Bolt_M4x20	
🗋 Procurement costs	0.15 EUR
▼ ☐ A-0000164 - LAMP_ver2_Bolt_M4x20	
🗋 Procurement costs	0.15 EUR
▼ ☐ A-0000127 - LAMP_ver2_QR_fromSTL	
🗋 Procurement costs	0.04 EUR
▼ ☐ A-0000124 - LAMP_ver2_Adapter	
🗋 Procurement costs	0.14 EUR
▼ ☐ A-0000148 - LAMP_ver2_Switch	
🗋 Procurement costs	6.63 EUR
▼ ☐ A-0000126 - LAMP_ver2_Lamp	
🗋 Procurement costs	5.16 EUR
▼ ☐ A-0000166 - LAMP_ver2_Bolt_M6x40x25	
🗋 Investment costs	0.35 EUR
▼ ☐ A-0000126 - LAMP_ver2_Lamp	
🗋 Porcurement costs	5.93 EUR

3D-lamp equipped with a standard light bulb that operates 8 h per day and is compared to the usage of a LED light bulb with the same utilisation. The assumed lifecycle amounts to one year.

The results of the comparison between the two light bulbs mainly focus on the differences in the investment and energy costs. The 566% higher investment costs of the LED light bulb are compensated by the massively reduced energy consumption and the associated energy costs, as visualised in Fig. 6.12.

The overall LCC of the LED 3D-lamp version are 41% lower than for the standard 3D lamp, thanks to the enormous energy cost savings. Figure 6.13 depicts and compares the life cycle cost of the two 3D-lamp versions and highlights the significant life cycle cost savings. Moreover, the higher investment costs of the LED 3D-lamp are amortised in only 23 days, as an indicator for the limited economic risk of the investment.

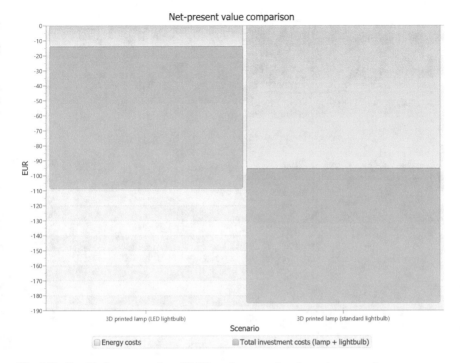

Fig. 6.11 Graphical representation of LCC results comparing alternatives scenarios

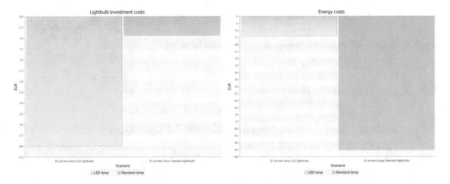

Fig. 6.12 Cost comparison of the 3D-lamp alternatives

In general, the LCPA approach can be adapted for the assessment of PSSs. Besides the relevant input data for products, like investment costs, energy costs, maintenance and other operating costs, the analysis of the service part requires input data that is more focused on, for instance, personnel costs, equipment usage to perform the service, travel costs, as well as service fees as revenues. As a result, the comparative LCPA approach enables the evaluation of different PSS concepts against each other.

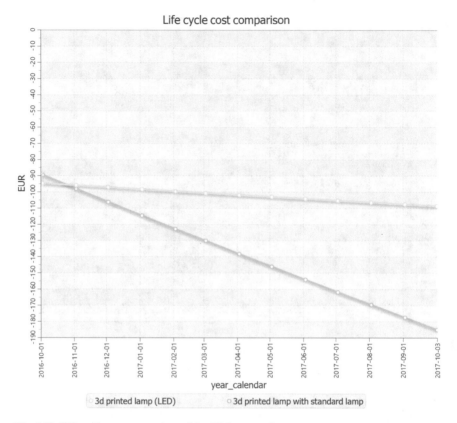

Fig. 6.13 Life cycle cost comparison of the 3D-lamp versions

6.9 Conclusion

The two tools developed for carrying out LCA and LCC analysis compliant with the Manutelligence needs have been introduced and their application to the assessment of a 3D-printed lamp has been described. The main advantage of these tools compared to commercial ones is their integration with the Manutelligence platform that allows a seamless use of the assessment results into the design process. Though any designer can benefit from the obtained results when comparing different alternatives, the assessment procedure and, in particular, the PSS modelling require some expertise in the field of LCA and LCC. Yet, both MaGA and BAL.LCPA tools pave the way for a more widespread use of LCA and LCC analysis for PSSs among practitioners.

References

1. Amaya J, Lelah A, Zwolinski P (2014) Design for intensified use in product–service systems using life-cycle analysis. J Eng Des 25(7–9):280–302
2. Bey N, McAloone TC (2006) From LCA to PSS–Making leaps towards sustainability by applying product/service-system thinking in product development. In: 13th CIRP International Conference on Life Cycle Engineering: LCE2006-Towards a Closed Loop Economy. pp 571–576
3. Corti D, Fontana A, Montorsi F (2016) Reference data architecture for PSSs Life Cycle Inventory. Procedia CIRP 47:300–304
4. Dal Lago M, Corti D, Wellsandt S (2017) Reinterpreting the LCA standard procedure for PSS. Procedia CIRP 64:73–78
5. Datta PP, Roy R (2010) Cost modelling techniques for availability type service support contracts: a literature review and empirical study. CIRP J Manufact Sci Technol 3(2):142–157
6. Doualle B, Medini K, Boucher X, Laforest V (2015) Investigating sustainability assessment methods of product-service systems. Procedia CIRP 30:161–166
7. ISO 14040 (2006) International standard, environmental management—life cycle assessment—principles and framework. International Organisation for Standardization, Geneva
8. ISO 14044 (2006) International standard, environmental management—life cycle assessment—requirements and guidelines. International Organisation for Standardization, Geneva
9. ISO 15686–5 (2008) Buildings and constructed assets—Service-life planning—Part 5: Life-cycle costing. International Organization for Standardization, Geneva
10. Kjaer LL, Pagoropoulos A, Schmidt JH, McAloone TC (2016) Challenges when evaluating product/service-systems through life cycle assessment. J Clean Prod 120:95–104
11. Kwak M, Kim H (2013) Economic and environmental impacts of product service lifetime: a life-cycle perspective. In: Product-service integration for sustainable solutions. Springer, Berlin, pp 177–189
12. Mannweiler C, Siener M, Aurich JC (2012). Lifecycle cost oriented evaluation and selection of product-service system variants. In: Proceedings of the 2nd CIRP IPS2 Conference 2010; 14–15 April; Linköping; Sweden (No. 077). Linköping University Electronic Press, pp 21–26
13. Matsumoto E, Ohtake J, Okada J, Takata S (2013) A method for selecting delivery modes in environmentally benign product service system design. In: The philosopher's stone for sustainability. Springer, Berlin, pp 375–380
14. Niemann J, Tichkiewitch S, Westkämper E (2009) Life cycle evaluation. In: Design of sustainable product life cycle. Springer Science & Business Media, pp 58–59
15. Ny H, Hallstedt S, Ericson Å (2013) A strategic approach for sustainable product service system development. In: CIRP design 2012. Springer, London, pp 427–436
16. Peruzzini M, Germani M (2013) Investigating the sustainability of product and product-service systems in the B2C industry. In: Product-service integration for sustainable solutions. Springer, Berlin, pp 421–434
17. Sakao T (2013) Engineering PSS (Product/Service Systems) toward sustainability: review of research. In: Handbook of sustainable engineering. Springer, Dordrecht, pp 597–613
18. Van Ostaeyen J, Kellens K, Van Horenbeek A, Duflou JR (2013) Quantifying the economic potential of a PSS: methodology and case study. In: The philosopher's stone for sustainability. Springer, Berlin, pp 523–528

Chapter 7
Use Cases

Claudio Violi, Laura Calvo Duarte, Catalina Amengual Garí, Pekka Puranen, Christian Norden and Lars Oscarsson

Abstract This chapter describes in details all the different use cases involved in the project. It is focused on how each of them has applied Manutelligence methods and tools in order to improve Product-Service system design and management. A particular focus is dedicated to the usage of the Manutelligence platform.

7.1 Ferrari Use Case

7.1.1 The Ferrari Company

Ferrari S.p.A. is an Italian luxury sports car designer and manufacturer based in Maranello. Founded by Enzo Ferrari in 1929, as Scuderia Ferrari, the company sponsored drivers and manufactured race cars before moving into production of

C. Violi (✉)
Ferrari S.p.A. Via dell' Abetone Inferiore, 4, 41053 Maranello, MO, Italy
e-mail: ClaudioAlfredo.Violi@ferrari.com

L. C. Duarte · C. A. Garí
Universitat Politècnica de Catalunya-BarcelonaTECH, Centre CIM, C/Llorens i Artigas, 12, 08028 Barcelona, Spain
e-mail: lcalvo@fundaciocim.org

C. A. Garí
e-mail: camengual@fundaciocim.org

P. Puranen
Meyer Turku Oy, Telakkakatu 1, 20240 Turku, Finland
e-mail: pekka.puranen@meyerturku.fi

C. Norden
BALance Technology Consulting GmbH, Contrescarpe 33, 28203 Bremen, Germany
e-mail: christian.norden@bal.eu

L. Oscarsson
Lindbäcks Bygg AB, Hammarvägen 21, 943 36 Öjebyn, Sweden
e-mail: lars.oscarsson@lindbacks.se

© The Author(s) 2019
L. Cattaneo and S. Terzi (eds.), *Models, Methods and Tools for Product Service Design*, PoliMI SpringerBriefs, https://doi.org/10.1007/978-3-319-95849-1_7

street-legal vehicles in 1947. Since then, Ferrari has caught on race tracks and roads all over the world more than 5000 victories, creating the basis of the Ferrari legend. Ferrari road cars are generally seen as a symbol of speed, luxury and wealth. The famous symbol of the Ferrari race team is the Cavallino Rampante ("prancing horse") black prancing stallion on a yellow shield, usually with the letters S F (for Scuderia Ferrari), with three stripes of green, white and red (the Italian national colors) at the top.

7.1.2   The Ferrari Business Challenges

The main challenges in developing Ferrari cars derive from the Ferrari corporate business drivers, which are:

- Deliver outstanding cars (innovative, high performance and reliability, cost controlled, enhances Product-Service).
- Deliver best product portfolio (manage product differentiation, exploit special series and supercars, high configuration offering, address traditional and new markets).
- Shorten time to market (high frequency of new car introduction, decrease risk of failing behind the market, control product development and manufacturing).

Obviously, these business drivers affect the whole vehicle life cycle from the concept to the after-sale; furthermore these drivers dictate extremely high rules that the enterprise has to fulfil towards their customers (product excellence, customer attitude and service). For these reasons the usual development and the manufacturing functions in Ferrari are very special. In detail, the process from concept to delivery is very long and sometime could take more than one year and a half. Thus, due to such a long period and complexity of processes, several challenges should be taken into account:

1. During the time horizon the technical characteristics as well as the specifics could change therefore they have to be integrated into the vehicle, which could be already entered from the engineering into the manufacturing process. Although this ensure to the customer the best of last technologies and some more features not yet available at the moment of the order, on the other hand it leads a complex process of requirements traceability and change.
2. The creation of a vehicle addresses disparate lifecycle phases, especially among design, engineering and manufacturing. One of the most important phases is the validation of the vehicle in term of design, this means that the virtual and physical prototype have to be benchmarked as easily and reliably as possible using dashboards and KPI. In addition, the resulting data from testing have to be used as feedback in the design and engineering phase in order to optimize the product. This triggers the need for fast, ubiquitous and secure sharing of product and service information across the entire Product-Service lifecycle involving all the relevant decision makers from the different functions.

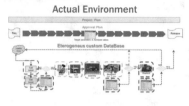

Fig. 7.1 The Ferrari objectives

3. The enterprise Ferrari would like to offer to its customers not only a car but also a strong experience that should start from the delivery of the vehicle or even more challenging without buying the vehicle. Thus, services related to the product should be designed in parallel and managed along the usage of the car as well as the related requirements.

7.1.3 The Ferrari Objectives in Manutelligence Project

The Manutelligence project offered the opportunity to evaluate how to address some of the previous challenges, especially the ones related with the Product-Service business. Effectively, if is it possible to retrieve data from physical prototypes during the tests and afterwards to provide services and feedbacks to the designers in the same way, changing the data model, it could be possible to collect data from the drivers experience and then to provide specific services to customers. This requires to adopt a platform able to support in a seamlessly way the access to all the needed information, from the usage of the car by the end users to the design and manufacturing project data (Fig. 7.1).

Business Objectives

- Improve the design of existing or new car, based on data coming from the real usage of the car.
- Captured data simulating a car customer testing on a circuit, to improve the design to obtain an enhanced "drive line" accuracy/driving comfort.
- The integration of the IoT information, to grab end user driving styles, technical data acquisition and elaboration, with the designer system tools in a single platform constitutes the innovative character providing a new way to design the Product/Service (Fig. 7.2).

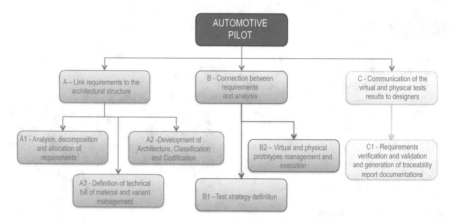

Fig. 7.2 Automotive pilot

Going more in detail, the following items have to be supported by the Manutelligence Platform:

- Management of product and service requirements and configuration features for variants connected.
- Define product template (archetype) and KBE.
- Link target parameters to the top level architecture.
- Definition of a complete, hybrid or partial product configuration, depending on the activity of test to perform, to be defined from beginning.
- Full traceability of produce and service requirements and accurate management of change impacts during design activities on design parameters to consider.
- Configure DMU to support physical and virtual test.
- Design validation process and optimization of product performance.
- Organize feedback to designers and data acquisition from field.
- To extend and improve the use of Simulation and optimize it through use of data collected from the field.

Business Scenario

In order to develop the Manutelligence platform a specific Ferrari Product-Service scenario has been identified. The FXX Programmes emerged out of the ingenious and rather fascinating idea of involving a group of special customers in the development of the Ferrari of the future, asking them to help provide information to the "Corse Clienti" technicians. Indeed, the enthusiasts who own these cars take part in a number of technical test sessions over the year closely monitored by Ferrari experts and have the chance to meet Maranello's engineers and professional testers in an environment in keeping with the tradition of the world's most famous race team (Fig. 7.3).

Innovative business model introduced in 2006
- Customers pay **annual fee** for:
  - Driving Course
  - 2 Full seasons (8 races)
  - Secured Factory Storage
  - Authenticity Certificate

SERVICES provided
- Event package:
  - driver pass
  - Hotel deluxe room
  - Technical advice, Physio therapist;
  - Coaching sessions with professional driver;

Fig. 7.3 FXX programme servitization-driven R&D

7.1.4 Use Case—Drive Style Comfort and Accuracy

In order to validate the Manutelligence platform a specific use case was selected, the Drive Style Comfort and Accuracy. The Drive Style is strongly depending on the accuracy, i.e. the capability to be precise while using the steering wheel to execute the best path, either on the circuit either on the road, and on the comfort, i.e. the easiness to control the steering wheel that requires to minimize any vibration deriving from the powertrain transmission system.

The objective is to improve the performance of the driving sensitivity and precision at low and high speeds by reducing the amplitude of the vibration of the steering line, caused either by the road irregularities (holes or kerb) either by the vibrations induced by the system itself. The vibrations cause a loss of sensitivity (sensitivity/driving accuracy) so that the driver is reducing speed when on circuit or is subject, when driving on the road, to a more or less evident fatigue driving.

Then the use case was analyzed executing one session of test on Fiorano circuit using FXX car and one specialist test driver, capturing the data related to the steering line, composed by the suspension corner, the steering box, the steering column and the steering wheel. In parallel the virtual model of the car was developed to execute frequency and modal analysis. Using the test data, the virtual model was calibrated (Figs. 7.4, 7.5, 7.6 and 7.7).

Another fundamental analysis was setup in order to enable the traceability of the tests. The management of the requirements, the target parameters, the test execution data, the BOM of the specific car used in the test, the test execution measurements were implemented. This means to manage

- Serialized BOM.
- Test case versus test results.
- Traceability matrix (components/tests data).

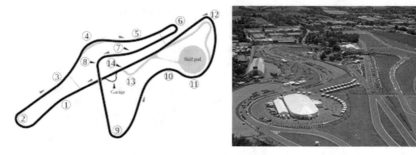

Fig. 7.4 Ferrari XX programme Fiorano circuit

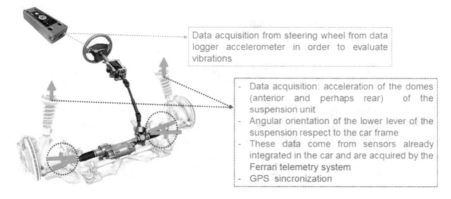

Fig. 7.5 Data acquisition from the steering line

Fig. 7.6 Data acquisition on the 3DEXPERIENCE platform

The final target is to improve the development of new car with a process that is able to capture and organize data coming from the real usage of the product by the end users, leveraging the IoT technology, to develop a virtual model to simulate the real usage in a quick and at less cost way, leveraging on the Design and Manufacturing platform solution. This iterative process will support the improvement of the Product-Service design in a more efficient and time-to-market oriented approach (Fig. 7.8).

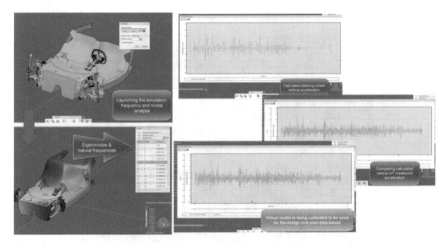

Fig. 7.7 Virtual model and data elaboration on the 3DEXPERIENCE platform

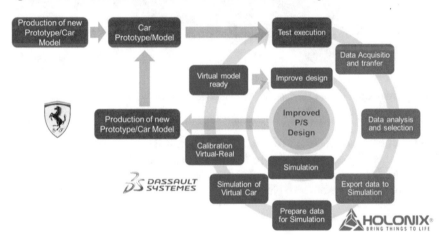

Fig. 7.8 Improved product and service design methodology

7.2   FabLab—Ateneus of Digital Fabrication (ADF)

7.2.1   The FabLab—Introduction

The city of Barcelona has created a FabLab facility with the MIT badge, promoted by the founder of the Institute of Advanced Architecture of Catalonia (IAAC). A Fabcity vision is being pushed through the gradual opening of public funded Ateneus

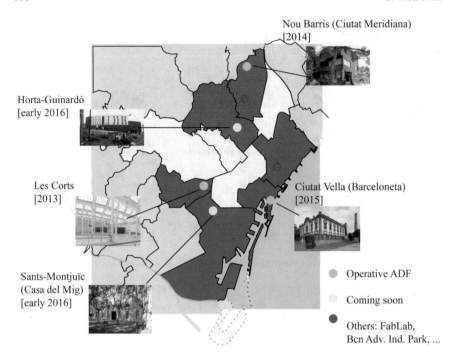

Fig. 7.9 Ateneus of Digital Fabrication (ADF) in Barcelona city

of Digital Fabrication (ADF) for each Barcelona district, as can be seen in Fig. 7.9. Besides, there are other FabLab type facilities like the Barcelona Advanced Industry Park. The network of FabLab/ADF is envisaged as becoming part of the public infrastructure of a sustainable city.

7.2.2 Context and Motivations

The use case FabLab/ADF aims at extending the Manutelligence concepts and tools to an emerging production paradigm for digital fabrication and rapid prototyping. The level of customization in a FabLab/ADF is usually very high, leading to a less structured design and production environment that have to meet the needs of customers with different level of expertise.

The implementation of five scenarios has been carried out for the FabLab/ADF pilot within ManuTelligence scope (Fig. 7.10).

Each scenario is divided in different use cases in order to cover all the process performed in a FAB-LAB.

In the next sections the different scenarios and use cases are explained and also the practical approach used during the project.

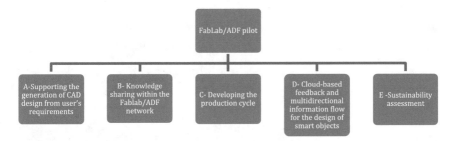

Fig. 7.10 FabLab five scenarios

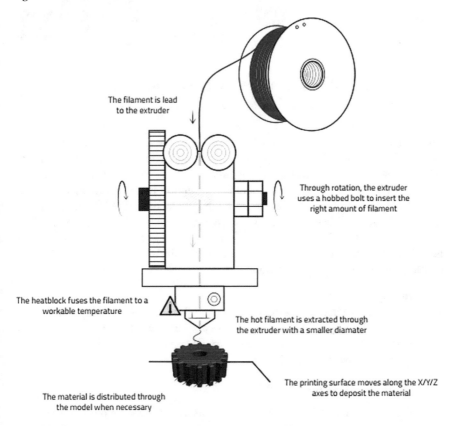

The filament is lead to the extruder

Through rotation, the extruder uses a hobbed bolt to insert the right amount of filament

The heatblock fuses the filament to a workable temperature

The hot filament is extracted through the extruder with a smaller diamater

The printing surface moves along the X/Y/Z axes to deposit the material

The material is distributed through the model when necessary

Fig. 7.11 Schematic representation of fused deposition modeling process. (Copyright BCN3D Technologies, image extracted from the User manual BCN3D+ page 4)

Fused Deposition Modelling (FDM) is the technology used for this practical implementation. FDM is a method of rapid prototyping that, starting with a digital 3D model divided on thin layers, consists on melting a filament—usually made of plastic—and deposition of those layers on a build platform (Fig. 7.11).

Fig. 7.12 Description of scenario A

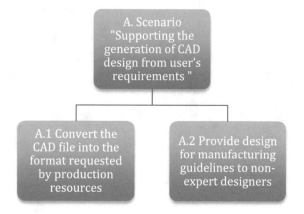

7.2.3 Scenario A: "Supporting the Generation of CAD Design from User's Requirements"

This scenario aims at improving the generation and management of CAD files (both 2D and 3D models), the first step towards the production of a prototype. In fact, several customers approaching the FabLab/ADF network are likely to be not so familiar with the use of a CAD and could need support in translating their idea into a CAD file, first, and in a format usable by the production resources after. For this reason, the Manutelligence platform is expected to make easier the generation and conversion of CAD file also by showing the users some design for manufacturing guidelines that can reduce the time it takes to FabLab/ADF operators to review and adjust a design developed by a customer.

Challenges in Scenario
The main challenge in the scenario was the identification of a set of suggestions that could be in some way automated within the platform and that are suitable for the additive manufacturing resources but also for other manufacturing technologies like laser cutting/engraving (Fig. 7.12).

Use Cases and Practical Approach

1. Convert the CAD file into the format requested by production resources.
2. Provide Design For Manufacturing guidelines to non-expert designers. In order to support the customer/user in the development of the CAD design, some guidelines in the form of Design For Manufacturing suggestions are provided by the platform, thus reducing the time it takes to FabLab/ADF operators to review CAD design and to make them producible (Fig. 7.13).

Fig. 7.13 Guidelines available on the platform

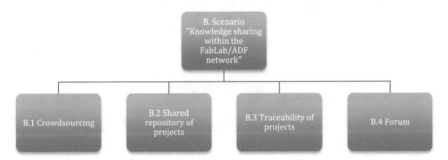

Fig. 7.14 Description of scenario B

7.2.4   Scenario B: "Knowledge Sharing Within the FabLab/ADF Network"

The objective of this scenario is to make easier the collaboration, communication and sharing of information between customers/designers who attend FabLab/ADF facilities to turn their ideas into real products.

The result of this scenario is also a process design that is more structured than the current one, thanks to the inclusion of the Bill of Material (BOM) of the products, representing a real innovation. Defining BOM of a product is an extended strategy for more complex industries such as car, ship or aerospace.

Challenges in Scenario

Integrate different technological platform in the scope of FabLabs/ADF networks (Fig. 7.14).

Use Cases and Practical Approach

1. Crowdsourcing. To involve more people in the realization of a project. The inclusion of a simplified BOM of a product will facilitate LCA and LCC analysis (see scenarios E). The possibility to add information in the platform by expert users and the possibility of using forums for the knowledge exchange facilitates collaboration between different disciplines.

Fig. 7.15 File versions generated in 3DExperience allows design traceability

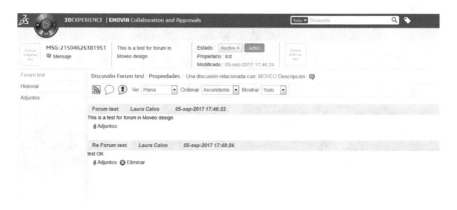

Fig. 7.16 Chat room test

2. Shared repository of projects. This use case seeks to implement an accessible portfolio of projects within 3DExperience software, adapting tools and strategies already implemented in more complex products and services (as the ones associated to cars, ships or aerospace).

This use case is planned for being a step-forward in the manner of sharing Fablab/ADF projects, typically represented by 3D printing products and services, currently supported by unconnected tools.

3. Traceability of projects. Closely connected to the previous point, this deployment will create a structure for the data repository of a single project so that it will possible to build a history of each product keeping trace of contributors, different versions, dates, production cycle information and so on (Fig. 7.15).
4. Forum. To make easier the communication of participants to the network, a tool that will enhance the possibility of sharing ideas, asking for support, providing expertise and creating joint projects, to name the main uses (Fig. 7.16).

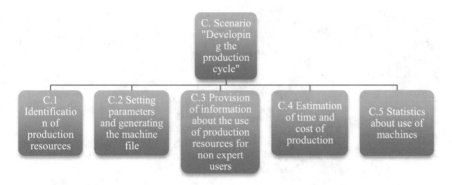

Fig. 7.17 Description of scenario C

7.2.5 Scenario C: "Developing the Production Cycle"

The Manutelligence platform is also expected to act as a link between the design phase and the production phase by making smoother the overall process to get the final product. This scenario focuses on the use of the platform as a support for the definition of the production cycle and for the analysis of some production parameters including time and cost.

Challenges in Scenario
The main challenge for implementing this scenario was related to data needed to make the system working. A lot of data about the use of machines and their performance had to be analyzed in order to populate the platform.

Use Cases and Practical Approach

1. Identification of production resources. From the analysis of the CAD design, the production resources that can be used for the production are identified (Fig. 7.17).
2. Setting parameters and generating the machine file (Fig. 7.18).
3. Provision of information about the use of production resources for non-expert users. Non-expert users can rely on some online information about the use of the selected machines, thus reducing the time it takes to FabLab/ADF operators to provide information about them (Fig. 7.19).
4. Estimation of time and cost of production. When the production cycle is defined, the customer/user is provided with the estimation of the production time and of the production cost (in terms of materials and energy).
   In order to achieve a better estimation of production time, Manutelligence project proposes to gather real data of the production machinery (Fig. 7.20). Then the cost estimated for the product will include the cost of all the energy really used, including for example the cost of producing scrap parts.

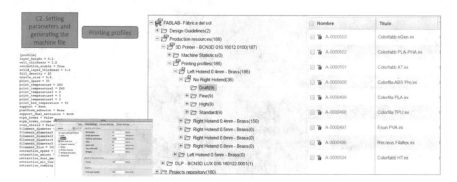

Fig. 7.18 Printing profiles added in the 3DEXPERIENCE

Fig. 7.19 Screenshot from the Platform

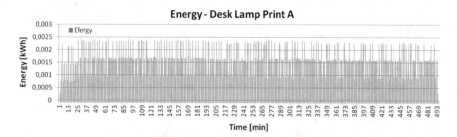

Fig. 7.20 Energy monitoring report

5. Statistics about use of machines. This pilot aims the monitoring of several digital production machinery included in a standard Fablab/ADF facility (3D printers, laser engravers, vinyl cutters…) with cost-affordable monitoring energy systems, capable of sending data to I-Like software from the Manutelligence platform and keeping it in a big database (Fig. 7.21).

Fig. 7.21 i-Like software from Holonix adapted to Fablab/ADF digital machinery

7.2.6   Scenario D: "Cloud-Based Feedback and Multidirectional Information Flow for the Design of Smart Objects"

The objective of this scenario is to set up an area of the ManuTelligence platform to be dedicated to collection and storing of data coming from sensors embedded in smart objects for carrying out multiple analysis. Users can have access to these data to analyse the product performance and behaviour during the use phase. On the other hand, FabLab/ADF operators could access the whole database to carry out statistical analysis on the smart objects with the aim of improving their expertise on them and being able to provide a better support for the design of this kind of product.

Challenges in Scenario

The main challenge was due to the innovativeness of this scenario compared to what happens at the moment: usually, products are not smart and during the use phase no information are collected. It means that training of FabLab/ADF operators who will

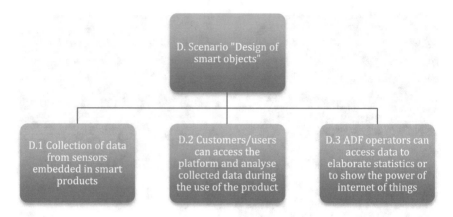

Fig. 7.22 Description of scenario D

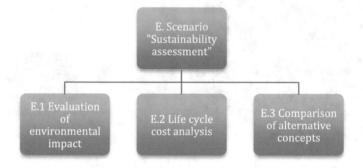

Fig. 7.23 Description of scenario E

be requested to become familiar with new tools and technologies is requested. On the other hand, also users need to be trained in order to correctly use sensors and perform analysis.

Use Cases and Practical Approach

1. Collection of data from sensors embedded in smart products. The ManuTel-ligence platform will have an area dedicated to the real-time collection and storing of data coming from sensors embedded in FabLab production resources and even smart products developed in the facilities (Fig. 7.22).
2. Customers/users can access the platform and analyse collected data during the production of the product.
3. FabLab/ADF. Operators can access data to elaborate statistics or to show the power of Internet of Things (IoT).

Fig. 7.24   3D printed lamp

7.2.7   Scenario E: "Sustainability Assessment"

The objective of this scenario is the integration of LCC and LCA tools into the design process of the FabLabs/ADF. Methods suitable for little structured design process like the one in FabLabs/ADF have to be developed starting from the Manutelligence tools. During the design process, it is possible to provide information about the environmental and the economic sustainability of the product being developed. On one hand, this allows the choice of the more sustainable solution among the available options thus improving the level of sustainability of the FabLab/ADF production. On the other hand, also users not directly interested in sustainability are made aware of the sustainability impact of their decisions thus increasing the level of knowledge and attention about this topic that, per se, is a social aim.

Challenges in Scenario
The main challenge in this case was related to the introduction of one more step into the design process that implies also the training of FabLab/ADF operators who had been be requested to become familiar with new tools and new concepts. The knowledge of these tools should be enough to support users/customers in their understanding.

Use Cases and Practical Approach

1. The evaluation of environmental impacts. It was done with the Manutelligence adapted LCA tool, MaGA. The LCA of the lamp was done in MaGa using the data collected when building the 3D printed lamp (Fig. 7.24) (more information in Chap. 6) (Figs. 7.23 and 7.25).

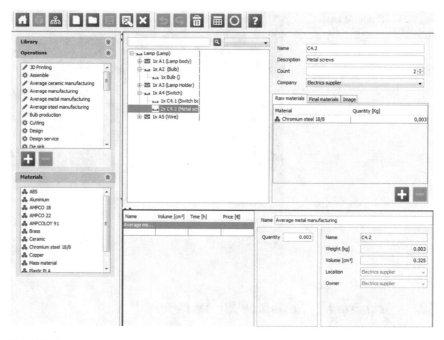

Fig. 7.25 Screen shot showing the creation of the 3D printed process in the MaGA software

2. Currently very simple models or methodologies are used to evaluate the cost of products and services developed within a Fablab/ADF, like using a table based on an excel sheet.

The aim of this pilot is to advance in evaluating the cost of developing a product and service within a Fablab/ADF. For this reason, the pilot is developing demo products such as the lamp for being implemented in LCPA software module by Balance. This use case is linked with the developments done in the Scenario C.

This way of evaluating the project developed within the Fablab/ADF helps to improve the evaluation of counterpart services that a customer of an Ateneu of Digital Fabrication (ADF) must do for using for free the digital production machinery installed for the city council.

3. Comparison of alternative concepts. The sustainability assessment provides data in (almost) real time so that it is possible to compare different concepts of the same product on the basis of its sustainability impacts. The designer is thus supported in the selection of the final design making him/her aware of the generated impact (more information in Chap. 6). An example is shown at Figs. 7.38 and 7.39.

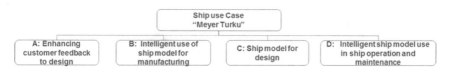

Fig. 7.26 Meyer Turku main scenarios in Manutelligence

7.3 Ship Use Case: Supporting Customer Feedback for Product-Service Design

7.3.1 Objectives and Results

The main goal is to move from product information management towards manufacturing intelligence, through more efficient management and integration of ship data and information. Focus is given on sharing the ship model to different stakeholders and supporting the communication between the different lifecycle phases, both forward and backward. This is needed to improve the engineering and service development, both cost- and time-efficiency and quality but also to create preparedness for offering the ship model as a service to the customer.

In Manutelligence, Meyer Turku currently defined four potential scenarios as described in Fig. 7.26 and several use cases belonging to them. From the beginning it was clear that not all of them could be implemented in Manutelligence. During the project the main focus was given to Scenario A: Enhancing customer feedback to design. The implementation of the scenario is described more in detail below.

The main objective in Meyer Turku use case is to receive feedback from the yard customer (ship owner) with new and more efficient communication methods. The idea is to add knowledge as a service to the product, to allow customer to view the ship with virtual reality and to be able to have more visual communication about current product prototype. In shipbuilding, product prototype means first ship in ship series. Normally one ship series contains 1–5 sister ships. With improved customer communication, it is possible to increase value of product by fulfilling customer needs in better way. Virtual reality tools are used as a communication method, because, from physically very large objects like a ship, it is difficult to create real 1:1 prototype or mock-ups. This idea can be also used in other product lifecycle phases, like between design and production. With new methods of communication it is possible to have more streamlined processes and finally to have better and more competitive product (Fig. 7.27).

Traditionally, material value is added in production to convert material into a product. In service sector, knowledge is sold as a service product. Combining product and service in the same entity as Product-Service makes better and more intelligent business.

In Meyer Turku use case, new service layer was added to standard shipbuilding process (shipbuilding process is basically a material management process).

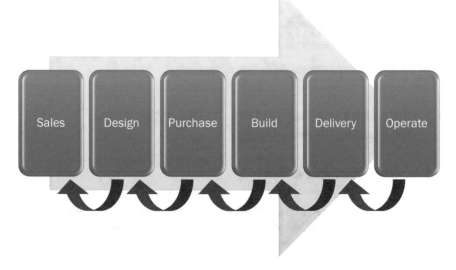

Fig. 7.27 Processes and feedback

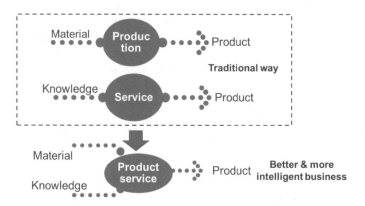

Fig. 7.28 The product-service combination

New process was developed for creating a virtual reality experience from Yard's standard 3D-models. See example below about example output in HTC-Vive virtual reality device (Fig. 7.28).

New process includes automatic clean-up of CADMATIC 3D model and own customized tools, which are created in Unity3D game engine (Figs. 7.29 and 7.30).

One example about customized tools is measurement and commenting tools inside virtual reality (Fig. 7.31).

Another example is the usage of MMO (Massive Multiplayer Online)-concept, where several engineers can connect to same virtual reality session by using Avatars.

Fig. 7.29 Example output in HTC-Vive virtual reality device

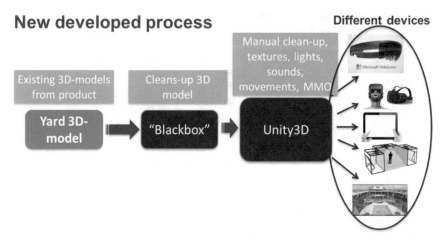

Fig. 7.30 Scheme of new developed process

Avatar characters can use different hand movements and also VoIP communication is possible. (Voice over Internet) (Fig. 7.32).

Meyer Turku has created in Manutelligence project demonstrator, which can be used as a collaborator-tool in real time usage. See picture Fig. 7.33 about idea and layout of demonstrator.

The parts of the demonstration include:

- Presenter has got Laptop computer or mobile phone, which is connected to MMO-cloud (Massive Multiplayer Online). Presenter can show own desktop with standard data projector.

Fig. 7.31 Example of comments into the virtual reality

Fig. 7.32 Meyer demonstrator: networked VR-platform mock-up

- Mobile VR-platform (virtual reality) where it is possible to see 3D-model with HMD-device (Head Mounted Device) via MMO-cloud. It is also possible to see "Presenter" and "Satellite-site in Finland" as avatar in virtual reality environment. Same information without HMD can be seen also in normal display in the table.
- Satellite-site in Finland is connected to same MMO-could and audience in the final project review can see and hear also what's happening in satellite-site in Finland.

With practical approach and by creating a real existing demonstrator it makes possible to really present new and more efficient communication layer to existing shipbuilding process.

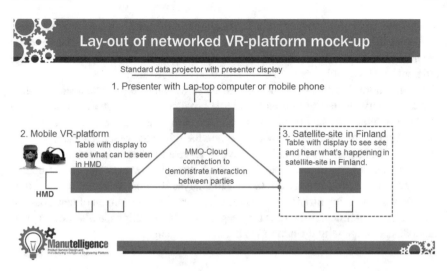

Fig. 7.33 Layout of networked VR-platform mock-up

7.4 Smart House Case—Introduction to Lindbäcks and Their Business Challenges

Lindbäcks is a family business from Piteå in northern Sweden. In close cooperation with our customers and with a huge motivator for innovation, Lindbäcks develops environmentally friendly and economical construction of apartment buildings in wood—all to create effective and sustainable places for tomorrow's needs. Lindbäcks place great emphasis on high quality accommodation where good materials and good design are the core ingredients. By always cooperate with some of the country's premier architectural offices, we create beautiful, healthy and practical homes where people enjoy and want to live.

Lindbäcks is Sweden's leading company in industrial construction of apartment buildings with a good knowledge of construction, production processes and project development. With the goal of becoming Europe's most modern producer of apartment buildings we are currently building a brand new production plant in Haraholmen, Piteå. It will be inaugurated by the end of 2017 and be fully powered by solar energy and district heating. Together with our production plant in Öjebyn, Piteå we will be able to produce 2400 apartments/year in 2020.

Lindbäcks houses are designed to last for more than a century. Based on the relatively long life cycle, the knowledge about the behaviour of the product throughout its lifecycle is crucial to remain competitive. Besides complaints during the warranty time, failure management and adjustments of new house orders from established customers, LINDBÄCKS receives limited feedback from the use-phase of their houses. In fact, it is a missed opportunity for iterative product improvements. Part of the required use-phase information is the monitoring of the energy consumption. While the theoretical energy consumption is calculated in the house design process, there is no real life validation of the actual energy consumption. Since the energy costs are becoming a cost driver in terms of life cycle costs, the real life proof of energy efficient houses could become a competitive advantage on the market.

Furthermore, Lindbäcks is looking for ways to enhance their business. Adding additional services to the houses to offer a product-service system is one way of thinking to approach new business areas. There are several ideas to establish additional services for living and how to integrate them into the product service system. However, the service for living ideas have to serve a market. Therefore, the service for living needs an appropriate business model, that relies on a precise cost structure. Internet of things technology could help to analyse the cost structure of the envisaged service for living offering and thereby support the development of new business areas. The chapter describes, how the usage of the Manutelligence platform enables Lindbäcks to cope with the addressed business challenges.

7.4.1 Collecting Information of the Use-Phase to Improve Lindbäcks Houses

To tackle the business challenges described in Chap. 1, Lindbäcks pursues the approach to implement internet of things technology within the Manutelligence project. The installation of the sensors is supposed to provide data to track the quality of Lindbäcks houses throughout the lifecycle and to gather the actual energy consumption. The gathered data of the sensors is transferred to the I-like platform of Holonix to serve as input parameters to the house designer at Lindbäcks and to the tenants, who can track his own energy consumption. Moreover, life cycle analysis based on the measurement data could serve as marketing instrument for Lindbäcks to position their houses as sustainable in the market (Fig. 7.34).

In a first step, Lindbäcks selected several key parameters to measure throughout the lifecycle.

Key parameters in the LINDBÄCKS use-case
Cold water
Water warm
Heating system
Humidity in wood
Humidity air
Temperature air
Water leakage kitchen sink
Energy consumption of floor heating systems

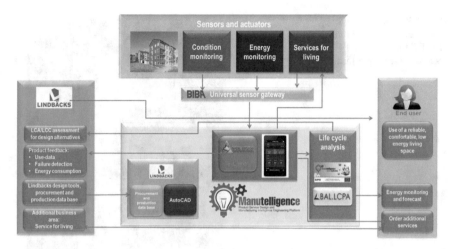

Fig. 7.34 LINDBÄCKS use-case approach in the Manutelligence project

Humidity in wooden houses is a key issue. Therefore, humidity sensors in the wood as well as in the room should be installed. Moreover, the room temperature shall be measured to identify causes for potential moisture issues and to crosslink the temperature information with the heating energy consumption. Sensors to detect water in the sink is conceived to be a prototype instalment to monitor water leakages in pipes, which can lead to major damages of the houses. In addition, the water consumption is measured to identify unusual water consumption combined with an alarm system to mitigate the risk of major water damages.

The measurement and sensor selection process followed a first prototype instalment in a flat of a Lindbäcks employee. Thereby the prototype consist of one air temperature sensor, one air humidity sensor and a sensor to detect water leakage in the kitchen sink. The gathered data has been collected by the universal sensor gateway unit and transferred to the I-LIKE module of the Manutelligence platform. The overall prototype testing underlined the potential benefit of developing and utilising the Manutelligence platform.

After the first prototype, a second, more comprehensive instalment has been planned. The second instalment comprises room temperature measurements in five different rooms. In addition, the outdoor temperature is measured as well. The measurement data is transferred to the I-Like-module of the Manutelligence platform, where that data values and data visualisation is handled (see Fig. 7.35). House designer and house operator can access the I-like module on a computer or a mobile device to analyse the data or to check the alarm.

In addition, the gathered data of the use-phase of a house enables to perform life cycle assessment and life cycle costing. To ease this process, a webservice has been established to import the required life cycle data into life cycle analysis tools, like BAL.LCPA. Based on the real measurements, the calculation of the operating

Fig. 7.35 Measurement value representation on the I-LIKE module of the Manutelligence platform

costs of the use-phase, especially costs for heating, is more precise and lead to more meaningful results (see Fig. 7.36).

The gathered data of the use-phase serves also as feedback -loop to Lindbäcks house designer. Based on the knowledge about the long use-phase of Lindbäcks houses, house designers can improve future house iterations and thereby improve the overall product quality in the long-term perspective.

7.4.2 Use-Phase Information to Create Additional Services

The collected information of the use-phase may also enable Lindbäcks to create new business areas. Developing additional services for living to be offered in combination with LINDBÄCKS houses could be realised based on the gathered knowledge. In the Lindbäcks use-case, the development of bathroom floor heating as a pay per use-service for the tenants has been developed. On the one hand, the technical realisation is based on state of the art technology. As part of the prototype testing described in Chap. 2, a bathroom floor has been equipped with a floor heating system as well as measurements of the energy consumption, system status information (on/off) and an

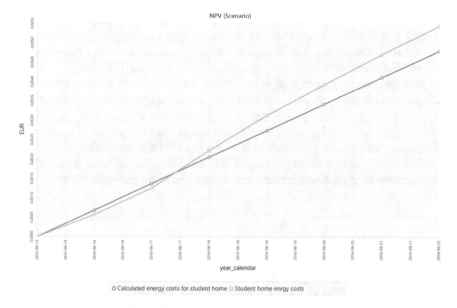

Fig. 7.36 Energy costs comparison theoretical values versus measured values

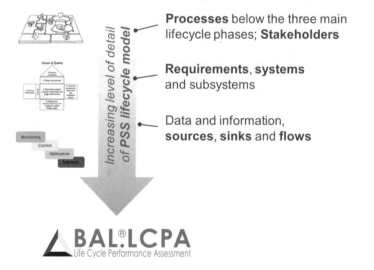

Processes below the three main lifecycle phases; **Stakeholders**

Requirements, systems and subsystems

Data and information, **sources, sinks** and **flows**

Increasing level of detail of PSS lifecycle model

BAL.LCPA
Life Cycle Performance Assessment

Fig. 7.37 Manutelligence PSS system development methodology

actuator to change the system status remotely. On the other hand, the development of the service including business model, pricing and payment options is more complex. Therefore, the Lindbäcks use-case applies the PSS system development of the Manutelligence project to create a new product service system (see Fig. 7.37).

The PSS development methodology starts with the development of a business model for the PSS in a business model canvas. In the second step, a quality function deployment is carried out. A third step includes an evaluation of the IoT functionality of the PSS. Concurrent to the first three steps, a life cycle model is developed containing relevant processes, stakeholders, and flows along the life cycle of the PSS. In a last step, the life cycle model is imported to life cycle analysis tools like BAL.LCPA, to evaluate the long-term profitability potential of the developed PSS. Part of the profitability analysis is the price definition of the pay-per-use service. Thereby, two different approaches for price finding are supported: On the one hand, a detailed life cycle cost analysis with profit margin can determine the price, or focus group testing among potential customers identifies the accepted price for the offered service.

7.4.3 Conclusion

The Manutelligence platform enables Lindbäcks to tackle their business challenge in terms of gathering product feedback throughout the lifecycle, optimising the actual energy consumption of their houses and developing new business areas. In particular, the internet of things implementation combined with the I-Like platform for data analysis and -visualisation creates valuable product feedback to improve Lindbäcks houses in the long-term perspective. Based on the gathered information of the use-phase, Lindbäcks expects to reduce material and construction failures, lower repair costs due to earlier recognition and optimise the energy efficiency of their houses. In addition, the life cycle analysis tools supports the product optimisation process and their results are supporting Lindbäcks in positioning their wooden houses as sustainable products in the market. The Manutelligence platform also encourages Lindbäcks to explore new business areas with its own methodology to develop product-service systems.

In fact, after the end of the Manutelligence project, Lindbäcks envisages to install IoT sensors in multiple commercial houses to gain the described benefits.

Chapter 8
Business Exploitation

Manuela Zacchei, Silvia Capato and Gianicola Loriga

Abstract Business based on Product-Services is growing within the manufacturing sector, where companies are increasingly improving the service component of their offering to gain competitive advantage. Actually, the aggregation of product and service lifecycle engineering tools is able to generate a significant added value and the analysis carried out in the following paragraphs is going to deeply illustrate the business potential of such tools within the manufacturing sector, with particular attention to the use cases of the Manutelligence project. The whole process (the aggregation and the value creation) will be therefore investigated and the starting point will be the analysis of Product Lifecycle Management, the relative technologies and the market associated. From the scenario analysis of the global PLM sector emerged that the aggregation of product and service lifecycle engineering tools is able to generate significant added value highlighting in this way the relevant commercial perspectives of the Manutelligence platform. Finally a business model for the Platform is presented.

8.1 Product/Service Life Cycle Management Scenario, Market Trends and Challenges

Insights into the global PLM market are given by CIMdata Inc., strategic consulting and research firm, which publishes every year global and regional analysis reports covering the Product Lifecycle Management (PLM) market. According to CIMdata's analysis, in year 2016, the PLM market grew to $40.6 billion overall, 5.0% growth in U.S. dollars over 2015. By analysing the specific PLM solution segments,

M. Zacchei (✉) · S. Capato · G. Loriga
Rina Consulting S.p.A. Materials, Technology & Innovation, Via San Nazaro, 19, 16145 Genoa, Italy
e-mail: manuela.zacchei@rina.org

S. Capato
e-mail: silvia.capato@rina.org

G. Loriga
e-mail: gianni.loriga@rina.org

L. Cattaneo and S. Terzi (eds.), *Models, Methods and Tools for Product Service Design*, PoliMI SpringerBriefs, https://doi.org/10.1007/978-3-319-95849-1_8

Electronic Design Automation (EDA) counts for about 20.6% ($8.3 billion) of the overall market, Systems Integrator/Reseller/VAR represents the 15.7% ($6.3 billion), cPDm (collaborative Product Definition management) is the third most important segment within the PLM market with a share 13.9% ($5.6 billion), S&A (Simulation & Analysis) counts for 13% ($5.2 billion) whereas Architectural, Engineering, and Construction (AEC) represents about the 8.3% ($3.3 billion). After the five most relevant segments, there are: Mechanical Computer-Aided Design (MCAD) Multi-Discipline counting for about 8.8% ($3.5 billion) of the market, MCAD focused representing about 6.5% ($2.6 billion), Focused Application Providers (including visualization and collaboration, content, document management, etc.) with a share of 5.1% ($2.0 billion), NC (Numerical Control) Non-Bundled with 3.1% ($1.2 billion) and Digital Manufacturing counting for 1.8% ($717 million).[1]

Most of the PLM market leaders presented a relevant growth in the last years and continued to make strategic acquisitions to expand their portfolios as well as in some cases, to enter new markets. In this framework, CIMdata forecasts the PLM market to grow at CAGR of 6.3% to $52.3 billion in 2020.[2]

Currently, in the PLM market among the solution providers there are four big players (Autodesk, Dassault Systèmes, PTC and Siemens PLM), Enterprise resource planning (ERP) vendors (Infor, Oracle, SAP) and others providers including new players offering different kind of PLM solutions among other products. The product innovation, product development and engineering processes result to be the priorities in today's vendors objectives. As far as the larger players are concerned, the PLM market is evolving on 4 key aspects: (1) Reaching more people (more customers, including also SMEs with cloud-based solutions); (2) Covering a richer view of the product; (3) Enabling more processes; (4) Supporting further up and down the product lifecycle.[3]

The current PLM solutions mostly rely on managing CAD models, documents, BOM, Product configurations and Simulations, more and more enlarging the scope to the manufacturing planning and management, with a minor coverage of the product lifecycle tasks following the delivery to customer.

Hence, business potential is envisaged for advanced PLM solutions able to manage all the processes related to the product lifecycle, including also the service component: there is the need to have a PLM solution that integrates in one platform every aspect of the entire product-service lifecycle.

More in details[4,5] the PLM market is going towards a significant transformation and the main trends are the following:

---

[1] https://www.cimdata.com/en/news/item/8281-cimdata-publishes-executive-plm-market-report (CIMdata 2017 Executive PLM Market Report).

[2] https://www.cimdata.com/en/news/item/6459-cimdata-publishes-plm-market-and-solution-provider-report (The CIMdata 2016 PLM Market and Solution Provider Analysis Report).

[3] http://tech-clarity.com/plm-vendors-2015/4269 (Strategies of the Major PLM Vendors 2015+).

[4] http://tech-clarity.com/strategies-of-the-major-plm-vendors-2016/5048 (Strategies of the Major PLM Vendors 2016+).

[5] https://blogs.oracle.com/PLM/entry/3_trends_driving_big_changes.

1. Platformization: there is a transition from standalone PLM, CAD, CAM, and CAE tools to Integrated Innovation Platforms able to provide a holistic, integrated approach enabling cross functional collaboration. There is the need of advanced capabilities in order to manage all the different phases and data related to the entire product lifecycle (i.e. from innovation and product development, to commercialization and service management).
2. Service Management: there are great business potential for commercialization of PLM tools that handle the service management across the whole product lifecycle, considering services integrated with the product-related data and activities.
3. Market dynamics changed by the cloud: there is relevant number of manufacturers and vendors that are beginning the transition to the Cloud.
4. Internet of Things (IoT): IoT and Industry 4.0 are dramatically reshaping the PLM sector.
5. Innovation platform: PLM solutions are evolving towards the support to bottom up innovation, providing companies with a systematic approach to capture, select and invest in the promising innovative ideas, allowing at the same time financial and cost analysis. PLM solutions are becoming a business strategy, helping companies to be competitive and innovative in their sectors.
6. Industry specialization: vendors are continuing their industry specialization.

8.2 Business Potential of the Manutelligence Platform

Product-Service Life Cycle Management (P-SLM) in fact enables manufacturing enterprises to drastically improve their development, production processes and delivery of personalized product-service, adding customer-focused innovative services. P-SLM takes into account the entire life cycle of a Product-Service, it strongly reduces new processes and plant designs time and allows planning and integrating new processes and procedures.

However, advanced tools, ICT platforms and methodologies are not enough. A dedicated business strategy is needed to maximise the exploitation potential of the Manutelligence solution, which will facilitate the shift from PLM to P-SLM.

After having analysed the reference PLM market and the current trends, the focus of this paragraph is therefore on the definition of the business potential enabling the Manutelligence consortium partners to create a shared business vision and to define their position in the market. This will facilitate their collaboration and different roles, and put the basis for the business and exploitation planning after the end of the project.

In particular, the business potential will be analysed at three different levels, which will correspond to three distinct exploitation strategies: (1) Platform developers; (2) Consulting business; (3) Specific use cases.

8.2.1 Platform Developers

The business strategy for the platform developers foresees the exploitation of both the overall Manutelligence platform and of the standalone modules.

The partners involved in the development of the platform and in the integrations of the different modules are several: Dassault Systemes, Biba, Balance, Holonix and SUPSI. The final platform will be based on Dassault 3D EXPERIENCE platform. This platform is already providing solutions for different industry sectors leveraging the different modules (brands) that offer embedded processes and functionalities (e.g. CATIA and Solidworks for 3D modelling or ENOVIA for collaboration and BOM management) In addition, different modules could be customised for each use case.

In particular, the modules developed in the project—such as the LCA/LCC (the MAGA module from SUPSI and the BAL.LCPA module from Balance respectively) and IoT modules—could either become central applications or interface with the platform as standalone tools or pure standalone without interfaces. In the latter case, end-users of the individual modules will not be able to use and manage all the data belonging to the 3D EXPERIENCE environment. The stand-alone modules will be owned by the single partners that developed them (and they could be separately sold by each developer) and a joint exploitation strategy will be defined by the end of the project by the different developers for the overall platform including all the modules.

The main users of the platform will be designers involved in the definition process of the product-service, marketing people and product managers that will be able to collaborate with each other as well as with the manufacturers all along the product development process. Anyhow, also the other actors engaged in the manufacturing value chain will be possible users of the platform, interacting and collaborating through that with the other players. Last but not least, the platform will allow following the maintenance product phase and will provide means to sustain the service business by collecting via IoT all the product usage information. The assessment software tool included into the platform, based on well-known methodologies of LCA and LCC, will support the modelling of the product service system and the collection of the huge amount of data, shortening the time occurring between a change in the design and the sustainability assessment.

The platform will provide an integrated environment for the management of all phases of product lifecycle, addressing a wide range of industries. Great business potential is envisaged for the exploitation of this platform in the manufacturing sector where the increased servitization, complexity of processes and use of cloud and big data require advanced instruments for the product-services lifecycles management. As a consequence, the business potential for Manutelligence software developers stands on the opportunity to integrate their own software, already used in several industrial sectors, into the overall Manutelligence platform, offering a competitive tool able to store all the information concerning the product lifecycle in a homogeneous way.

Initially, the platform will be based on a generic concept of product-service lifecycle and this could represent a weakness of the product for those industrial sectors demanding specific requirements. However, following adaptations and improvements of the platform may be implemented in order to create more customizable solutions and enhance their usability in specific sectors according to the trends underlined in the previous paragraph. Considering that currently, in the market, none of the existing tools provides functionalities tailored to PSS, with environmental and economic evaluation integrated in the same solution, the value added of Manutelligence platform entails great market perspectives.

8.2.2 Consulting Business

Manutelligence consulting business companies can exploit the Manutelligence platform as a mean to support industrial customers in process/product development as well as by offering them training on the management of product lifecycle and knowledge based on advanced S-PLM technologies.

The Manutelligence platform helps increasing the quality of interaction with customers and third parties, besides improving collaboration between different business areas inside the company itself. Quality of the services provided will be enhanced because the platform allows industrial clients to develop innovation within their businesses, providing them with full traceability of all data related to product lifecycle stages. The market trends of the manufacturing sector (i.e. servitization, evolution towards global market, etc.) work as opportunities for the exploitation of the platform by consulting companies. Several industrial companies in fact need support for complex product development activities, whereas product-related services have to be managed in relation to the product lifecycle. Even more important, manufacturing companies need to find innovative solutions to differentiate from competitors and defend/acquire market share. Overall, the targeted customers will be small-medium manufacturers (SMEs), aiming to reach the global market in the consumer goods, automotive, home & architecture, civil engineering, energy, automation and packaging sectors. All these sectors, for the most part, are not mature yet for a complete understanding and utilization of S-PLM technologies, thus they need to be trained before being interested direct buyers for Manutelligence platform. Ideally, the target customers have low development rate (products change every 2/3 years) or small portfolio of products, a not fully structured technical office and they are fostering an innovative approach to design. Among the Manutelligence partners usually providing consulting services, it is worth to mention Rina Consulting (formerly D'Appolonia), VTT, BIBA, Politecnico di Milano and Balance.

8.2.3 Uses Cases

The users of the Product-Service Lifecycle Management system (i.e. Ferrari, Meyer Turku Oy, Lindbäcks and Fundacio CIM) will have the opportunity to unlock a significant market potential and gain a clear competitive edge through the development of customized and knowledge intensive products and services.

Automotive

The Automotive use case is conducted at Ferrari facilities. Ferrari is leader of this pilot while other beneficiaries, namely Politecnico di Milano and Dassault, are involved with the aim of scientific, methodological and technical support.

Ferrari represents the luxury segment of the automotive sector and its business has a strong focus on the delivery of innovative product-services, in order to offer a customer experience based on the "thrill of driving".

The whole automotive industry has been experiencing an evolution towards servitization. In the last decade, automotive companies faced the economic crisis exploiting revenues deriving from services linked to the purchase of a vehicle. Hence, manufacturers and dealers continue to expand the service offering, providing clients with customizable services across all phases of the product lifecycle. Therefore, the automotive sector deals with both important technical and "experience" aspects, but it is still centred on product development activities. Firms need new business models based on a product-service concept, besides advanced software tools able to manage big amount of data related to products' lifecycles increasingly complex. Therefore, a huge business potential could be unlocked by the Manutelligence platform within the automotive sector. Manufacturers will benefit from a tool that can be integrated with existing software to perform, in addition to other more traditional tasks, also sustainability assessments through LCA and LCC methodologies and the management of all customers' data. More specifically, automotive manufacturers will have the competitive advantage to guarantee the customers with the best technology and technical specifications available at the time of delivery and not at the time of the order, since status of the car under manufacturing will be traceable at each moment of its lifecycle (3D virtual representation) and advancements occurring between design and delivery stages could be implemented. Other consequent benefits rely on reduction of the time before delivery, better quality of product-services provided, thanks to clients' feedback data, and increased cost savings which is a high priority for manufacturers. In fact, Manutelligence platform will allow data from testing and use of the car to be feedback into the design and engineering phases in a standardized manner in order to optimise the product ordered.

Shipbuilding

The Shipbuilding pilot is led by Meyer Turku Oy, one of the leading European shipbuilding companies. The company provides state-of-the-art technology solutions, advanced construction processes and innovations for cruise operators and other ship owners. The cruise sector is based on strong branding/marketing strategies aimed at always attracting new customers. Cruise operators are challenged to develop competitive cruise packages involving a high-quality stay onboard, an array of shore-based activities offering access to a variety of cultures and sites and easy transfers to/from the vessel. The ship itself is designed as a tourist attraction.[6] As a matter of fact, to fulfil the desires of its guest and attract new customers, the cruise industry has started diversifying the cruise product, to be able to respond to the preferences of a wide range of customer groups. To this purpose, the industry has innovated through the development of new destinations, new ship designs, new and diverse onboard amenities, facilities and services. Also for this reason, cruise liners are among the largest and most complex products made. Such complicated and interconnected systems hold a great market potential for the Manutelligence platform. Shipbuilding companies will be able to benefit from a tool that supports and enhances the interactions between engineering, ship manufacturing and construction, customers and users and which could also be integrated with existing and new software.

Smart House
The Smart house pilot is focused on the combination of modular housing components with sensing and communication technology, for the design and construction of future student homes. The pilot leader is Lindbäcks. The development and the construction of healthy homes indoor, which are then assembled quickly on site, have a high potential. The objective of this pilot is to demonstrate that room designers and manufacturers can provide personalised housing solutions—both product and services—to student residents. The data of usage gathered and the sensors measurement will in fact give valuable feedback to design and production departments. Moreover, additional apartment features will help the construction company gain a competitive advantage. Considering the trends in the overall construction industry, a great business potential will also be exploited with the introduction of additional per-use services, thanks to the installation of several sensors. These services include:

7. Service for living the apartment, for example adjust apartment setting like room temperature by voice, control of pets, older people, burglar alarm;
8. Service for maintenance, including sensors for humidity, control-system for water, heat and air consumption;
9. Service for construction, such as temperature, humidity, air pressure, etc.

In addition, other pay-per-use services could be investigated in the smart house use case to create an opportunity for additional business area. Pay per use services include: speaker in the kitchen, electrical heating floor, window with darken function, order kitchen equipment when needed, laundry by tube, carpool and bike pool in basement, pay rent after use of apartment. These services could be integrated in product offering, creating the opportunity to address customers' requirements across the entire lifecycle of the project. As a matter of fact, student residents will be able to adapt the rooms to their preferences and needs and be involved in the design phase.

Nevertheless, aspects related to data protection will have to be investigated to allow a full exploitation of the Manutelligence platform.

---

[6]J. Brida and S. Zapata, "Cruise Tourism: Economic, Socio-Cultural and Environmental Impacts," International Journal of Leisure and Tourism Marketing, vol. 1, no. 3, 2010.

Fab Lab

The FabLab use case is focused on FabLab-like facilities called Ateneus of Digital Fabrication (ADF), promoted by the Barcelona city council within the strategic Smart Cities framework. Fab Ateneus is a space dedicated to creation and learning connected to social innovation, new technologies and, especially, digital fabrication, where citizens are the active users and protagonists.

The objective of this use case is to give to Fab Ateneus users the possibility to learn and start using the potential of 3D printing and the "Internet of Things". In particular, the aim is twofold; first of all to enable collaborative design for FabLabs, allowing "makers" to cooperate using best in class design collaboration tools. Secondly, to enable the manufacturing of IoT enabled objects, through the possibility of adding single boards PCs with sensors to a fully working IoT platform. This will support the growth of the future generation of designers and engineers, used to think from the beginning to product-services and Internet enabled things. As above mentioned, Fab Ateneus exists to promote activities and projects to improve society (i.e. the neighbourhood, the city and the world) using new models of organization (co-creation, open source collaboration, crowdsourcing, crowdfounding, etc.) and open, networked learning and social media to share acquired knowledge. The Manutelligence platform will help "makers" undertake projects that will have a positive impact on the neighbourhood, the city and the world providing a platform that:

10. Supports the generation of CAD design from user's requirements, making easier the conversion and generation of CAD design from ideas just even sketched in a paper into readable format;
11. Enhances knowledge sharing within the FabLab/ADF network, with the creation of a network that allows the sharing of data about product design leading to a more efficient design process and to a better service for those customers who are not skilled;
12. Develops the production cycle, from the CAD file to the production resources needed for the realisation of the product, for the different digital fabrication technologies;
13. Gets cloud-based feedback and multidirectional information flow for smart objects;
14. Allows sustainability assessment. Environmental and economic sustainability factors will be included in the design phase to increase the level of awareness of FabLab/ADF customers and create more sustainable products, with less environmental impacts.

More generally, the Manutelligence platform will enable the transformation of the knowledge characteristic of the Information society into new economics forms centred on people, exploiting an important social vector.

8.3 Business Model

In this paragraph, a Business Model for the Manutelligence platform is presented. First of all, before developing the Business Model, a value proposition analysis has been conducted for each Manutelligence pilot (Automotive, Shipbuilding, Smart House, FabLab). Then, the outcomes of each value proposition analysis have been generalized in order to develop the business model of the Manutelligence platform. An overview of the Business Model Canvas resulted from the analysis is reported in Fig. 8.1.

8.3.1 Customers Segments

The analysis conducted for the definition of the Business Model Canvas started from the analysis of the potential customers and of the problems they are currently facing that are solved by the Manutelligence platform. Starting from the analysis of the pilots and their needs, the aim was to identify those ones that are cross to different sectors.

First of all, the users have been identified:

15. Designers: person that is responsible of the realization of a 2D/3D model.
16. Project managers: the main responsible of the definition and execution of project activities coordinating people involved and verifying progresses.

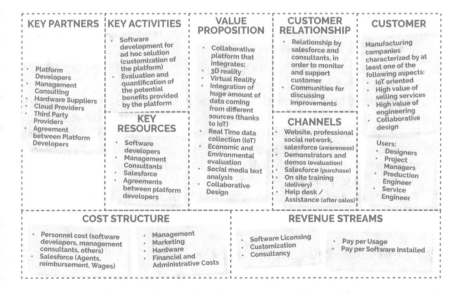

Fig. 8.1 Manutelligence business model canvas

17. Product engineers: responsible to define and control design activities and related resources as well as of the realization of the product.
18. Service engineers: responsible to define and control service activities and related resources as well as of to the realization of the service.

Furthermore, a list of characteristics of companies potentially interested in Manutelligence platform has been listed:

19. IoT oriented: manufacturing companies with the objective of the development of new smart products, provide new services to their customers, collect high amount of data from product or sensors, or gather real time information from machines.
20. High value of selling services: companies characterized by the creation of a large part of revenues from the selling of services.
21. High value of engineering: companies characterized by intense engineering activities where knowledge from previous project became essential to decrease the impact of the engineering within the overall cost of the developed solution. This requirement is typically requested by Aerospace, Aeronautical, Naval, Major Infrastructure, Oil & Gas industries.
22. Collaborative Design: companies interested on the improvement of the collaboration between different domain experts. This is typical from the already mentioned companies however this aspect could be crucial also for Mechanical, Software Houses, Creative, FabLab companies.

8.3.2 Value Proposition

The value proposition of the Manutelligence platform is the proposal of a collaborative platform that integrates different features, such as:

23. Physical and Virtual experience and information, integrated in the 3DEXPERIENCE of Dassault Systemes.
24. Integration of huge amount of data coming from different sources (thanks to IoT), integrated in the IoT Module of Holonix and Real Time data collection.
25. Economic and Environmental evaluation, integrated in the BAL.LCPA module (economic evaluation) and in the MAGA module (environmental evaluation), leveraging on interface with 3DEXPERIENCE of Dassault.
26. Social media text analysis of BIBA.
27. Collaborative design, integrated in the 3DEXPERIENCE of Dassault.

8.3.3 Customer Relationships and Channels

The different ways to reach customers creating awareness, selling and delivery of the Manutelligence platform have been identified. Concerning the creation of awareness, the development of an "ad hoc" website, the advertising on different professional social network (such as Linkedin) and the salesforce have been identified how effective streams for reaching the potential customers. In order to enable the customers to evaluate the potentialities and the benefits of the Manutelligence platform, the idea is to develop some demonstration and demos that show how the platform works. Salesforce is used to sell the platform to the interested customers, providing also consultancy and training activities (on site). The relation with customers after the sales is managed mainly through a helpdesk, based on a website platform, or through a direct assistance.

8.3.4   Revenue Models

In this area possible ways to gain revenues have been identified. A first way is the sell software licenses. Secondly, it is possible to create revenues through the consultancy and the customization of the platform, according to the customer's needs. Furthermore, it is possible to consider the Manutelligence platform as a service, therefore using other ways of revenues, such as the pay per usage (the customer pays just when it uses the platform) or pay per software installed (the customer pays only for the software it needs).

8.3.5   Key Activities and Resources

The key activities to make the Manutelligence platform a commercial solution ready for the industrial environment can be summarized in:

28. Software development for ad hoc solution (customization of the platform), in order to satisfy the customers' needs.
29. Evaluation and quantification of the potential benefits provided by the platform through precise and measurable KPIs.

The key resources to provide the platform on the market are:

30. The solution architect, in order to finalize the platform design for the daily business activities.
31. The software developers, in order to implement the platform according to the solution architect design.
32. The management consultants, in order to support customers in the choice of the proper platform.
33. The salesforce, in order to facilitate the communication of benefits and advantages reachable through the platform.
34. Agreements between platform developers in order to provide the platform on the market and realize some pilots/demonstrators.

8.3.6   Key Partners

Key relationships have to be established in order to facilitate the market introduction of the Manutelligence platform. First of all, an ad hoc agreement between platform developers is necessary to be signed in order to share revenues and investments. This is crucial for the joint exploitation of the project results. Clearly, the key partners will be the platform developers and an important role can be labelled by management consulting. External consortium partners, such as hardware suppliers, cloud providers and third party providers, are necessary to provide a complete solution, composed of the software (developed within the Manutelligence project) and of the hardware (provided by external partners).

8.3.7   Cost Structure

The cost structure distinguishes the periodic and the variable costs. Personnel costs cover a significant part within the whole costs. It is composed of the solution architects, the software developers, the management consultants and other costs. Another important part of the cost is composed of the salesforce, composed of agents' salaries, reimbursements and wages. Finally the management have a significant cost to coordinate the work giving a strategic vision. Other costs can be summarized in: (1) Marketing Costs; (2) Hardware; (3) Financial Expenses; Administrative Costs.

References

1. International Labour Organization "Construction sector" [Online]. Available http://www.ilo.o rg/global/industries-and-sectors/construction/lang–en/index.htm. Accessed November 2015
2. Osterwalder A, Pigneur Y (2010) Business model generation: a handbook for visionaries, game changers, and challengers. Wiley
3. Osterwalder A, Pigneur Y, Bernarda G, Smith A (2014) Value proposition design: how to create products and services customers want. Wiley
4. https://www.scottish-enterprise.com/~/media/se/resources/documents/stuv/sustainable-busin ess-models.pdf
5. Vosgien T (2015) Model-based system engineering enabling design-analysis data integration in digital design environments: application to collaborative aeronautics simulation-based design process and turbojet integration studies. Ecole Centrale Paris
6. Barkay J (2014) Service lifecycle managemen. Siemens PLM Software
7. McKinsey & Company (2013) The road to 2020 and beyond: what's driving the global automotive industry?
8. Lay G (2014) Servitization in industry. Springer International Publishing
9. Rodrigue J, Notteboom T (2013) The geography of cruise shipping: itineraries. Capacity deployment and ports of call, 3rd edn. Routledge, New York
10. Business Wire (2014) The global prefabricated buildings market—key trends and opportunities to 2017
11. Menichinelli M (2014) Come construire un FabLab. Design

Printed in the United States
By Bookmasters